MANIFESTATION 5- SCIENCE OF VIBRATIONS & FREQUENCIES

Kanav Sachdev

White Light Publication

ISBN-13:978-81-992617-2-3 (Hardcover)
ISBN-13:978-81-992617-7-8 (Paperback)
ISBN-13:978-81-992617-6-1 (Ebook)

Cover design by: Jyoti Sachdeva

To all the seekers, mystics, and curious souls who stood at the edge of the known and dared to ask "What if?"—this book is for you. For my spiritual teachers, past and present, whose wisdom helped illuminate the path. For my family, whose unwavering support allowed me to chase cosmic questions. And for those who feel unseen or unheard: may these pages help you discover the vibrational power within and manifest the reality your heart truly desires.

CONTENTS

Title Page
Copyright
Dedication
Introduction
Opening Note
Disclaimer 1
WHY THIS BOOK EXISTS 3
The Vibrational Universe 7
Chapter 1: Everything Vibrates 8
Chapter 2: The Frequency of Emotions 13
The Science of Sound 61
Chapter 3: Nada Yoga & The Power of Om 62
Chapter 4: Modern Sound Science 81
Chapter 5: Solfeggio & Healing Frequencies 92
Tuning Your World 137
Chapter 6: Vibrational Architecture 138
Chapter 7: Vibrational Carriers 150
Daily Rituals 166
Chapter 8: Morning Frequency Routine 167
Chapter 9: Micro-Resets During the Day 209
Chapter 10: Evening Frequency Reset 233

Mastery & Manifestation	260
Chapter 11: Aligning Desire with Frequency	261
Chapter 12: Living a High-Vibe Life	285
Chapter 13: The Science of Miracles	317
THE COMPLETE LIBRARY OF KANAV SACHDEV	332
THE ARCHIVES OF CONSCIOUSNESS	340
STANDALONE TITLES FOR SUCCESS	355
FOR BREAKING LAZINESS & INERTIA	357
FOR A TOTAL IDENTITY REBOOT	359
FOR CRUSHING ANXIETY & STRESS	361
FOR INSTANT CLARITY & INTUITION	363
THE QUANTUM SHORTCUTS	365
FOR URGENT FINANCIAL GROWTH	367
A Personal Note from the Author	369

INTRODUCTION

I spent twelve years studying manifestation systems. Every major method—visualization, affirmation, intention-setting, vision boards, energy work, ritual, meditation. I read the bestsellers. I attended workshops. I practiced with discipline. And yet, despite everything I understood intellectually about how manifestation *should* work, my financial situation remained stuck.

Not dramatically poor. But perpetually constrained. Every time I approached a threshold of real abundance, something would shift. A client would leave. An expected payment would delay. An opportunity would materialize and then dissolve. It was as if an invisible ceiling existed just above my head, and no amount of manifestation technique could lift me through it.

I watched others manifest wealth seemingly effortlessly. People with less knowledge, fewer credentials, less sophisticated understanding of spirituality and consciousness. They would speak something into existence and it would appear. They would set an intention and watch it materialize within weeks. Meanwhile, I was doing everything "right"—meditation, affirmations, visualization—and yet the universe seemed to have a different plan for me.

The frustration wasn't just about money. It was about the contradiction. How could I teach others about manifestation when I couldn't manifest consistent financial abundance in my own life? How could I write about frequency and vibration when my personal broadcast seemed misaligned with everything I was consciously intending?

That's when the real understanding shifted.

I was at a crossroads—reviewing research on neuroscience, specifically the HeartMath Institute's work on heart coherence. I had read about it before, acknowledged it intellectually, and moved on. But this time something landed differently. The research showed something stunning: that the electromagnetic field generated by the heart (which is 5,000 times stronger than the brain's field) creates distinct patterns based on emotional states. Chaotic, incoherent patterns when stressed. Organized, coherent patterns when genuinely grateful or at peace.

And then the realization hit with the force of a physical strike:

My conscious mind was broadcasting abundance affirmations while my nervous system was broadcasting financial fear. My brain was saying "I deserve wealth" while my body was saying "I'm not safe with money." The universe doesn't hear your words. It responds to your coherence—the alignment between what your mind states and what your nervous system broadcasts.

I wasn't experiencing a manifestation failure. I was experiencing a coherence failure.

That moment changed everything. Because suddenly, the problem wasn't that manifestation doesn't work. The problem was that I had been trying to manifest from a divided state. My conscious intention and my subconscious nervous system were broadcasting contradictory frequencies—and the universe, functioning perfectly as a responsive vibrational system, was giving me back exactly what my total broadcast was transmitting.

The work then became simple, though not easy: alignment.

Not more affirmations. Not stronger visualization. Not bigger vision boards. Instead, I began the actual work of making my entire being—mind, heart, nervous system, body—coherent around financial abundance. Not through positive thinking, but through genuine nervous system recalibration.

Within two months, the ceiling lifted. The first sign was subtle—a client returned after a year of no contact with a substantial project. But more importantly, the stuckness shifted. It was as if I had finally tuned myself to the frequency of someone who could receive and hold abundance. And once I was at that frequency, opportunities that had always been present suddenly became visible to me. Connections that had been dormant activated. Conversations that had been impossible became natural.

Not through magic. Through alignment.

This is what this book is about.

Not another technique to add to your toolkit. Not more affirmations or visualizations or law of attraction methods. This book is about understanding that manifestation has nothing to do with the force of your will or the sophistication of your technique. It has everything to do with the coherence of your vibrational broadcast.

Most manifestation books teach you to think differently. This book teaches you to become different—to shift the fundamental frequency you're broadcasting so that you naturally attract experiences that match it. Not through effort. Through alignment.

The science is clear. Your body generates an electromagnetic field. That field changes based on your emotional state. That field broadcasts into the universe. The universe responds to the frequency you're actually broadcasting, not the frequency you think you're broadcasting.

If you've tried everything and still not seen the results you want, this book offers the missing piece. Not because you were doing something wrong, but because you were broadcasting from a divided house. Your conscious desires were at one frequency, and your nervous system was at another. The universe received the truth—the total broadcast—and responded perfectly.

This book teaches you to become whole. To align every part of yourself—conscious and unconscious, mind and body, intention and nervous system—around what you want to manifest. Not as a technique, but as a complete recalibration of who you are at a vibrational level.

OPENING NOTE

Have you ever wondered why some people seem to effortlessly attract opportunities while others struggle despite relentless effort? My journey began from the same questions, countless experiments, and a fair share of failures. This book is the culmination of years spent searching for the invisible laws that shape our lives—the science behind why some dreams materialize and others vanish into thin air. What you hold in your hands is not another collection of feel-good affirmations, but a practical guide to the reality-shifting power of vibration and frequency. By integrating ancient Eastern wisdom, cutting-edge research, and direct personal experience, I invite you to explore what truly determines what you receive—and become the conscious designer of your life's symphony.

DISCLAIMER

Please read this disclaimer carefully before using this book or applying any of the principles described herein.

Fictionalization and Privacy

The case studies, anecdotes, and personal stories presented within this book are included for educational and illustrative purposes only. While they are inspired by real-world concepts and experiences, they are fictionalized. In all instances, the names, identifying details, professions, locations, and other key details of individuals have been altered or entirely fabricated to protect privacy and ensure anonymity. Any resemblance to actual persons, living or dead, or actual events is purely coincidental.

Not Professional Advice

The content of **Science of Vibrations & Frequencies** is intended to provide information, spiritual concepts, and personal development principles based on the author's interpretation of ancient texts and practices. It is not intended to be a substitute for professional medical advice, diagnosis, or treatment.

- Consult a Professional: Always seek the advice of your physician or other qualified health provider with any questions you may have regarding a medical condition or before undertaking any new breathing or physical practice, particularly those involving breath retention or physical implements (such as the Yoga Danda).
- Use at Your Own Risk: The practices described in this book should be approached with respect and caution. The

author and publisher are not responsible for any adverse effects or consequences resulting from the use of any suggestions, practices, or preparations mentioned in this book. Your engagement with the material is entirely at your own risk.

WHY THIS BOOK EXISTS

This book exists because there is a gap—a chasm, really—between what has been understood in ancient wisdom traditions for thousands of years and how most people approach manifestation today.

For millennia, Vedic sages understood something fundamental: that consciousness and vibration are not separate from physical reality. They are the foundation of it. The Upanishads teach that the universe arises from sound vibration—the cosmic Om. Kashmir Shaivism describes "spanda," the primordial throb that creates and sustains all existence. These weren't poetic metaphors. They were descriptions of how reality actually functions at its deepest level.

Modern quantum physics has confirmed what these ancient traditions knew: everything in existence is fundamentally vibrating energy. What appears as solid matter is actually energy condensed to specific frequencies. Your emotions, your thoughts, your beliefs—all exist as vibrations that interact with the vibrational field around you.

Yet this understanding has been largely absent from contemporary manifestation teaching.

Most manifestation books fall into one of two categories. They either teach from a purely spiritual perspective—affirmations, visualization, energy work—without explaining the actual mechanism of how it works. Or they teach from a purely scientific perspective—neuroscience, psychology, behavioral change—without acknowledging that consciousness itself is a vibrational phenomenon that operates according to its own laws.

This book bridges that gap.

It is grounded in both cutting-edge neuroscience (HeartMath research, brainwave entrainment, nervous system recalibration) and authentic ancient wisdom (Vedic principles, mantra science, frequency physics). Not as a curiosity. Not as comparison. But as integrated understanding—showing that the mechanisms described in the oldest spiritual texts are the same mechanisms being validated in modern laboratories.

More importantly, this book focuses on what most manifestation systems miss entirely: the difference between your conscious intention and your nervous system broadcast.

Your conscious mind can intend abundance. But if your nervous system is in a state of scarcity, fear, or unworthiness, the universe receives that broader transmission. You are broadcasting a contradiction. And the universe, which is fundamentally a responsive vibrational system, gives you back exactly what your total broadcast is transmitting.

This book teaches you to eliminate that contradiction.

Not through more willpower. Not through stronger

affirmations. Not through better visualization techniques. Instead, through genuine nervous system recalibration—becoming someone whose entire being, from conscious intention to subconscious broadcast, is coherent around what you want to manifest.

This is why the results come so quickly once alignment occurs. You are not forcing anything. You are not trying to manifest from a state of internal division. You are simply becoming someone to whom the desired reality naturally flows.

Every chapter contains:

The Science — Why this practice works, explained through neuroscience, quantum physics, or Vedic principles. You are not asked to believe anything. You are shown how it works.

The Practice — Step-by-step protocols you can implement immediately. Most take 5-20 minutes. Most can be done anywhere—at work, home, in traffic, in difficult moments.

Real Results — Stories from readers like you who implemented these practices and experienced specific, measurable manifestations.

The structure of this book moves from understanding (what frequency is and how it works) through application (how to shift your frequency) to integration (how to make this your normal, automatic way of being in the world).

By the end, you will have more than knowledge. You will have a complete system for becoming a conscious frequency transmitter—someone whose vibrational broadcast is so coherent and aligned with your desires that manifestation becomes not something you practice, but something you are.

This book exists because you deserve to know how reality actually works. Not theory. Not inspiration. The actual operating instructions of consciousness and vibration. And more importantly, you deserve a practical path to aligning your entire being so that what you want to manifest flows to you naturally.

That path begins on the next page.

THE VIBRATIONAL UNIVERSE

CHAPTER 1: EVERYTHING VIBRATES

The Hidden Code of Reality

What if the invisible difference between your struggling neighbor and your thriving neighbor isn't intelligence, effort, or luck—but vibration?

What if you could measure that vibration, understand it, and deliberately shift it to align with what you want?

In this chapter, you'll discover that everything in existence vibrates at specific frequencies—and yours can be consciously recalibrated. This pulsation isn't just happening within you. It's the fundamental nature of everything that exists.

Modern quantum field theory tells us something remarkable: what appears as solid matter is actually energy vibrating at specific frequencies. The chair you're sitting on, the book in your hands, even the thoughts flowing through your mind—all are expressions of vibrating energy fields, dancing in patterns too rapid for our eyes to perceive.

"All matter is merely energy condensed to a slow vibration," as physicist Albert Einstein once noted. This isn't just poetic language; it's a description of how reality actually functions at its most fundamental level.

What's fascinating is that this scientific understanding mirrors what ancient wisdom traditions have taught for thousands of years. In Kashmir Shaivism, one of the world's oldest spiritual philosophies, there exists the doctrine of "spanda"—the primordial vibration or throb that creates and sustains the universe.

The ancient yogis didn't have electron microscopes or particle accelerators, yet through deep meditation, they perceived this vibrational nature of reality. They understood that the universe isn't made of static, separate objects but is instead a unified field of pulsating energy.

The Upanishads, sacred texts from India dating back thousands of years, describe the universe as arising from sound vibration—the cosmic Om. Similarly, the Bible begins with, "In the beginning was the Word," suggesting that sound vibration preceded physical creation.

These aren't just interesting parallels between science and spirituality. They point to a profound truth about the nature of reality that has direct implications for how manifestation works.

If everything is vibration, then manifestation isn't about forcing change through willpower alone. It's about understanding and aligning with vibrational laws. It's about becoming a conscious participant in the vibrational dance of creation rather than struggling against it.

Think of it this way: you can't tune a radio to 94.5 FM and expect to hear a broadcast from 107.3 FM. The frequencies don't match. Similarly, you can't vibrate at the frequency of lack and limitation while expecting to receive abundance and opportunity. The universe responds not to your words or even your conscious thoughts, but to your vibrational signature.

This vibrational understanding changes everything about how we approach manifestation. It moves us from trying to manipulate external reality through force to aligning our internal frequency with what we wish to experience.

The code of reality isn't written in some secret language only accessible to the privileged few. It's written in the language of vibration—a language your entire being already speaks fluently, whether you're consciously aware of it or not.

Tesla's Whisper

"If you want to find the secrets of the universe, think in terms of energy, frequency, and vibration."

These words, whispered across time from the brilliant mind of Nikola Tesla, might be the most important clue ever given about how reality operates. Tesla wasn't just one of history's greatest inventors—the father of alternating current electricity that powers our modern world—he was a visionary who perceived the vibrational nature of existence decades before quantum physics confirmed it.

What makes Tesla's insight so powerful is that it came from someone firmly rooted in the scientific method, not from mystical speculation. This was a man who revolutionized our understanding of electricity and electromagnetic energy, who created technologies we still use today. When Tesla spoke about energy, frequency, and vibration as the "secrets of the universe," he wasn't speaking metaphorically—he was describing what he understood to be the literal operating system of reality.

Tesla's work with resonance and frequency led to discoveries that seemed almost magical to his contemporaries. He created oscillators that could match the resonant frequency of objects, famously claiming he could bring down buildings with a pocket-sized device by finding their resonant frequency. While that particular claim remains unverified, the principle behind it—that everything has a natural frequency at which it vibrates most powerfully—has been thoroughly proven.

This principle of resonance is central to understanding manifestation. When two systems vibrate at similar frequencies, they begin to vibrate in harmony—a phenomenon called "entrainment." We see this when pendulum clocks mounted on the same wall eventually swing in perfect synchronization, or when women living together find their menstrual cycles aligning over time.

The same principle applies to manifestation. When your personal vibrational frequency matches the frequency of what you desire, you naturally draw that experience toward you. It's not magic—it's physics.

Tesla understood that these principles weren't just applicable to mechanical or electrical systems but to all aspects of reality. He once said, "The day science begins to study non-physical phenomena, it will make more progress in one decade than in all the previous centuries of its existence."

Today, we're seeing Tesla's prediction come true as quantum physics explores the observer effect—how consciousness itself influences physical reality at the quantum level. What Tesla intuited through his work with electromagnetic energy, modern science is now confirming through rigorous experimentation.

For you as someone seeking to manifest your desires, Tesla's insight offers a practical approach: rather than focusing solely on actions or even thoughts, pay attention to your energetic state—your frequency. Learn to sense it, measure it, and most importantly, deliberately shift it.

This isn't about abandoning practical action. Tesla himself was incredibly productive, obtaining nearly 300 patents worldwide. Rather, it's about understanding that your vibrational state is the foundation upon which all effective action is built.

When you approach manifestation from this perspective, you stop fighting against the current of creation and start flowing with it. You recognize that the universe isn't a cold, mechanical system to be conquered but a vibrational symphony you're already part of—and one where you can learn to play your instrument with greater skill and harmony.

Tesla's whisper across time reminds us that the most powerful forces in the universe are invisible to the naked eye but perceptible to those who understand energy, frequency, and vibration. This book will help you become one of those people.

CHAPTER 2: THE FREQUENCY OF EMOTIONS

Map of Consciousness

Right now, your nervous system is broadcasting a completely different frequency than your conscious mind. You're sending contradictory signals to the universe.

This contradiction is why affirmations fail, visualizations don't manifest, and you feel stuck despite your best efforts.

This chapter teaches you to identify this gap and close it—permanently.

This isn't just a metaphor—it's a measurable reality that has profound implications for your ability to manifest what you desire.

Dr. David Hawkins, a psychiatrist and spiritual teacher with a background in physics, developed what he called the "Map of Consciousness"—a scale that assigns numerical values to different emotional states based on their vibrational frequency. While we won't focus on the specific numbers, understanding the hierarchy of emotions can transform how you approach manifestation.

At the bottom of the scale sit emotions like shame, guilt, and apathy. These emotions vibrate at the lowest frequencies and severely limit your manifestation power. When you're in these states, you're essentially broadcasting a signal that attracts more experiences that match these low vibrations.

Moving up the scale, we encounter fear, anger, and pride. While these emotions contain more energy than the bottom-tier emotions, they still create significant resistance to positive manifestation. Think about it—have you ever tried to manifest something wonderful while feeling afraid or angry? The results are usually disappointing because your emotional frequency contradicts what you're trying to create.

The middle of the scale includes courage, neutrality, and willingness. These states begin to open the door to effective manifestation because they contain less resistance. When you feel courageous enough to pursue your desires despite uncertainty, you're vibrating at a frequency that allows more possibilities to enter your experience.

At the upper levels of the scale, we find acceptance, reason, love, joy, and peace. These high-frequency emotions dramatically amplify your manifestation abilities. When you're in a state of genuine gratitude or love, you're broadcasting a powerful signal that attracts experiences matching these elevated frequencies.

At the very top sits enlightenment—a state of pure consciousness beyond ordinary emotion that few humans have sustained for extended periods. But you don't need to reach enlightenment to manifest effectively. Even moving from fear to courage, or from anger to acceptance, can create dramatic shifts in your results.

What makes this map so valuable for manifestation work is that it helps you understand why your results might not match your intentions. You might be visualizing abundance while feeling subtle anxiety about money—and that anxiety is broadcasting a frequency that contradicts your conscious desires.

The good news is that you can deliberately shift your emotional frequency. Unlike your height or eye color, your emotional state isn't fixed—it's fluid and responsive to your choices. With practice, you can learn to recognize when you're in a low-frequency state and use specific techniques to elevate your vibration.

This isn't about suppressing "negative" emotions or pretending to feel positive when you don't. Such fakery actually creates internal dissonance that further complicates your vibrational state. Instead, it's about honestly acknowledging where you are on the map and then using proven methods to genuinely shift to higher frequencies.

Think of it like tuning a radio. If you want to hear a station broadcasting at 101.5 FM, you need to tune your receiver to that frequency. Similarly, if you want to manifest experiences that vibrate at the frequency of abundance, joy, or love, you need to tune your emotional state to match those frequencies.

The Map of Consciousness isn't just a theoretical model—it's a practical tool that helps you understand why some manifestation efforts succeed while others fail. By becoming aware of your emotional frequency and learning to elevate it, you gain access to manifestation abilities you might never have realized you possessed.

HeartMath & Coherence

What if I told you that your heart generates an electromagnetic field that extends several feet beyond your body—and that this field changes shape depending on your emotional state? What if this field could be measured, graphed, and deliberately altered to enhance your manifestation abilities?

This isn't speculation. It's the groundbreaking work of the HeartMath Institute, a research organization that has spent decades studying the relationship between the heart, brain, emotions, and energy fields.

Using sophisticated equipment, HeartMath researchers have documented how different emotional states create distinct patterns in the heart's electromagnetic field. When you experience stress, frustration, or anxiety, your heart rhythm becomes erratic and disordered, creating what scientists call an "incoherent" pattern. This incoherence doesn't just stay inside your body—it radiates outward, affecting your energy field and, by extension, your ability to manifest.

Conversely, when you experience positive emotions like appreciation, care, or love, your heart rhythm becomes ordered and harmonious—a state called "coherence." This coherent pattern creates a strong, stable electromagnetic field that extends beyond your body and interacts with the fields of others and the environment around you.

Why does this matter for manifestation? Because your heart's electromagnetic field is 5,000 times stronger than the field generated by your brain. It's the most powerful rhythmic electromagnetic field produced by your body—a literal broadcast of your emotional frequency into the universe.

When your heart is in coherence, several remarkable things happen:

- Your brain synchronizes with your heart, creating whole-body coherence
- Your stress hormones decrease while beneficial hormones increase
- Your immune system functions more efficiently
- Your cognitive abilities sharpen
- Your intuition becomes more accessible

Most importantly for manifestation, heart coherence creates a vibrational state that's highly conducive to attracting experiences that match your desires. It's like clearing static from a radio signal—suddenly, your broadcast becomes clear, strong, and effective.

The HeartMath Institute has developed a simple technique called the Quick Coherence® Technique that allows you to shift into heart coherence in just a minute or two:

1. **Heart Focus**: Place your attention on the area of your heart in the center of your chest.
2. **Heart Breathing:** Breathe deeply but normally, imagining that your breath is flowing in and out through your heart area.
3. **Heart Feeling**: Recall a positive feeling—a time when you felt good inside—and try to re-experience that feeling now.

This simple practice has been shown to rapidly shift heart rhythm patterns from incoherent to coherent, changing your electromagnetic broadcast in real-time.

What's particularly valuable about the HeartMath research is that it provides scientific validation for what many spiritual traditions have taught: that the heart is more than just a pump—it's an intelligent organ capable of influencing your entire being and your relationship with the world around you.

Their research has shown that people in a state of heart coherence demonstrate improved manifestation abilities, including better problem-solving, increased synchronicities, and more favorable outcomes in their endeavors. This isn't mystical thinking—it's measurable science.

By learning to generate and maintain heart coherence, you're not just feeling better emotionally—you're literally changing your vibrational broadcast to the universe. You're creating an energetic state that's optimally aligned with successful manifestation.

The bridge between emotion and manifestation isn't just conceptual—it's electromagnetic. Your feelings create measurable energy fields that interact with the world around you in ways that either support or hinder your manifestation efforts.

Heart coherence gives you a scientifically validated tool to consciously participate in this energetic conversation, allowing you to broadcast the frequency of what you wish to manifest rather than the frequency of what you wish to avoid.

Practical Reset Tools

Life happens. Even with the best intentions, you'll encounter situations that pull your emotional frequency down. The key to successful manifestation isn't avoiding these moments—it's having reliable tools to quickly reset your vibration when they occur.

Here are seven practical techniques you can use anywhere, anytime to shift your emotional frequency and get back into alignment with what you want to manifest:

The Gratitude Burst

When your frequency drops, stop whatever you're doing and rapid-fire list five things you're genuinely grateful for right now. Don't just name them—feel them. "I'm grateful for the warmth of the sun on my skin. I'm grateful for the breath in my lungs. I'm grateful for the memory of my child's laughter this morning."

This works because gratitude and negative emotions cannot coexist in the same energetic space. The moment you activate genuine gratitude, you automatically shift your vibrational frequency upward.

Practice: The Gratitude Burst

Time Required: 2 3 Minutes

Best For: Rapid emotional frequency shift, when stuck in low vibe

Frequency: As needed, can use 3 5 times daily

Equipment Needed: None

BASIC VERSION (2 minutes)

1. Stop whatever you're doing immediately

2. Name 5 things you're genuinely grateful for RIGHT NOW

3. Don't just list them FEEL each one for 20 seconds

4. Example gratitude: "I'm grateful for the warmth of sun on my skin. I'm grateful for the breath in my lungs. I'm grateful for my child's laughter this morning. I'm grateful for this moment of choice. I'm grateful for my body's sensitivity to energy."

5. Notice the shift in your body shoulders relaxing, chest opening

ENHANCED VERSION (5 minutes)

After listing 5 gratitudes, add 3 MORE for things that were previously

obstacles:

- "I'm grateful for the financial struggle because it taught me discipline.
- I'm grateful for that heartbreak because it taught me self love.
- I'm grateful for that failure because it taught me resilience."

This deeper gratitude transforms blocks into bless

ings and elevates frequency even more powerfully.

SUCCESS INDICATORS

Physical Signs: Shoulders relax, breathing slows, chest opens, warmth spreads through body

Mental Signs: Mind becomes quieter, shift from problem focused to opportunity focused thinking

Energetic Signs: Lightness in chest, expanded heart space, feeling more resourceful and capable

Timeline: Immediate effect (within 30 seconds), intensifies over 2 3 minutes

TROUBLESHOOTING

If "I can't feel genuine gratitude right now":

Start with small things (breath, your hand, temperature).Gratitude begins small.

If "I feel forced or fake":

You're correct this won't work. Identify ONE thing genuinely good in this moment, no matter how small. Authenticity matters more than quantity.

If "Nothing shifts":

You might need the Enhanced Version or a different reset tool (try The Laughter Trigger instead).

INTEGRATION

Use BEFORE: Any manifestation visualization work

Use AFTER: Difficult experiences or triggering events

Combine WITH: The 3 Breath Reset for powerful 5 minute rebalancing

Don't combine with: Meditation on your fears/blocks (save gratitude for after processing emotions)

◆ ◆ ◆

The Laughter Trigger

Keep a collection of videos, memes, or memories that reliably make you laugh. When you feel your frequency dropping, take a 60-second laughter break. Even forced laughter triggers the same physiological benefits as genuine laughter, releasing endorphins and shifting your energy almost instantly.

A belly laugh is one of the fastest ways to elevate your vibration from a low state to a significantly higher one. Your body doesn't distinguish between laughing at something funny and laughing for no reason at all—the vibrational shift happens either way.

Practice: The Laughter Trigger

Time Required: 1 2 minutes

Best For: Rapid mood elevation, breaking negative thought loops

Frequency: As needed, 1 3 times daily when stuck

Equipment Needed: Access to funny video/meme/memory

BASIC VERSION (1 2 minutes)

1. Keep a collection of videos, memes, or memories that reliably make you laugh

2. When you feel your frequency dropping, access one immediately

3. Watch/recall it and allow yourself to laugh genuinely

4. Let the laughter continue for 30 60 seconds

5. Notice the immediate shift in your energy Store these on your phone or in a folder on your computer for instant access when needed.

ENHANCED VERSION (2 3 minutes)

Add intentional laughter even when nothing is funny:

1. Stand or sit in an open posture

2. Begin laughing intentionally don't wait for something funny

3. Let your body move with the laughter

4. Continue for 60 90 seconds

5. Your body won't distinguish between "real" and "forced" laughter both trigger the same neuro chemical release

This is actually MORE powerful than waiting for something funny because you're actively choosing the frequency shift rather than depending on exter nal stimuli.

SUCCESS INDICATORS

Physical Signs: Face muscles relax, abdominal ten sion releases,breathing becomes rhythmic and easy, body feels lighter

Mental Signs: Worry and anxiety dissolve, mind be comes playful,perspective shifts

Energetic Signs: Energy becomes buoyant and expan sive, sense of humor and lightness return

Timeline: Immediate (within 30 seconds of genuine laughter),residual effect lasts 15 minutes.

TROUBLESHOOTING

If "I can't find anything funny":

Try the Enhanced Version with intentional laugh ter. You'll be amazed at how quickly it becomes genuine.

If "I don't have funny videos prepared":

Go to your phone right now and create a folder of 3 5 funny videos before you need this tool. Prep aration is key.

If "Laughter doesn't work for me":

You might respond better to The Physical Pattern Interrupt or The Laughter Trigger combined with movement.

INTEGRATION

Use BEFORE: Creative work (laughter opens creative pathways)

Use AFTER: Stressful situations or bad news

Combine WITH: Dancing or movement for amplified energy shift

Great for: Breaking stuck thought patterns, emergency frequency resets

◆ ◆ ◆

The Physical Pattern Interrupt

Your body and energy field are intimately connected. When you change your physical state dramatically, your emotional frequency shifts in response. Try these pattern interrupts:

- Jump up and down while shaking your limbs for 30 seconds
- Dance wildly to an upbeat song
- Strike a "power pose" with your arms raised in a V above your head
- Splash cold water on your face

These physical changes break the pattern of low-frequency emotions and create space for a vibrational reset.

Practice: The Physical Pattern Interrupt

Time Required: 30 60 seconds

Best For: Breaking entrenched negative patterns, activating stuck energy

Frequency: As needed, especially useful when other resets aren't working

Equipment Needed: Space to move (can be minimal)

BASIC VERSION (30-45 seconds)

Choose ONE of these and do it right now:

1. ****Jump and Shake**:** Jump up and down while shaking your limbs rapidly for 30 seconds. Let your body move wildly and freely.

2. ****Power Pose**:** Stand tall with arms raised in a V above your head. Hold for 30 seconds while breathing deeply. Feel your personal power expanding.

3. ****Dance Wildly**:** Put on an upbeat song and dance without any concern for how it looks. Move ment breaks patterns faster than anything else.

4. ****Cold Exposure**:** Splash cold water on your face or hold ice cubes in your hands. The shock re sets your nervous system immediately.

5. ****Dynamic Stretching**:** Do large, exaggerated stretches of your entire body, reaching arms up, bending side to side, twisting your torso.

ENHANCED VERSION (1 2 minutes)

Combine movements for more powerful effect:

1. Jump and shake for 30 seconds (20 jumps)

2. Do 10 full body stretches reaching toward the ceiling

3. Spin in a circle 3 times

4. Do 10 power poses with arms in the air

5. Dance to one upbeat song

This full sequence creates a complete nervous system reset that can break even deeply entrenched low frequency states.

SUCCESS INDICATORS

Physical Signs: Heart rate increases, body feels warm, tension releases, muscles feel activated, breathing deepens

Mental Signs: Mental chatter quiets, sense of aliveness returns, mood lifts noticeably

Energetic Signs: Stuck energy releases and flows, sense of lightness and freedom, blockages dissolve

Timeline: Immediate effect (within 30 seconds), lasting 20 30 minutes

TROUBLESHOOTING

If "I feel self conscious doing this":

Go into your bathroom or a private space where no one can see.Privacy allows you to fully commit.

If "I don't feel like moving":

This is often when you MOST need this practice. Start with one small movement and momentum will build.

If "I don't have space to move":

You only need 3 feet of space. Or use the cold water splash or power pose variations that need minimal space.

INTEGRATION

Use BEFORE: Creative or manifestation work (acti vates energy)

Use AFTER: Sitting for long periods (restores flow)

Combine WITH: 3 Breath Reset afterward for calm activation

Most effective for: People who are kinesthetic/ physical/movement oriented

◆◆◆

The 4-7-8 Breath

This powerful breathing technique activates your parasympathetic nervous system, shifting you out of stress and into calm:

- Inhale quietly through your nose for 4 counts
- Hold your breath for 7 counts
- Exhale completely through your mouth with a whoosh sound for 8 counts.

Repeat three times

This technique works by altering your oxygen/carbon dioxide balance, triggering a relaxation response that elevates your vibrational frequency.

Practice: The 4 7 8 Breath (Box Breath ing Variant)

Time Required: 2 3 minutes

Best For: Anxiety relief, nervous system regulation, instant calm

Frequency: As needed, up to 3 times daily, best before bed

Equipment Needed: None

BASIC VERSION (2 minutes 3 full rounds)

1. Sit or lie comfortably with spine straight
2. Close your eyes (optional but helpful)
3. Exhale completely through your mouth

Then repeat the 4 7 8 pattern 3 times:

4. Inhale quietly through your nose for 4 counts
5. Hold your breath for 7 counts (count slowly)
6. Exhale completely through your mouth with a whoosh sound for 8 counts (make this exhale long)
7. Pause briefly at the end of the exhale
8. Repeat the pattern 2 more times (3 rounds total)That's it. Your nervous system will shift noticeably.

ENHANCED VERSION (3 5 minutes 6 full rounds)

Same pattern as Basic, but:

1. Do 6 complete rounds instead of 3
2. On each exhale, visualize releasing stress as grey smoke
3. With each inhale, visualize drawing in calm blue light
4. Count more slowly on the hold (stretch the 7 counts a bit)
5. Make the exhales even longer if possible (extend the 8 counts)

The extended practice creates deeper nervous system regulation and can be used when you need more pro found calm.

SUCCESS INDICATORS

Physical Signs: Heart rate slows, shoulders relax, jaw unclenches, breath becomes natural and easy, body feels heavy and grounded

Mental Signs: Racing thoughts slow, mental chatter quiets, mind becomes clearer, anxiety dissolves

Energetic Signs: Nervous system shifts from sym pathetic (fight or flight) to parasympathetic (rest and digest), sense of safety increases

Timeline: Noticeable effect within 1 2 minutes, strong effect by 3 minutes, lasting 20 60 minutes

TROUBLESHOOTING

If "I can't hold my breath for 7 counts":

Start with 2 3 4 or 3 4 5 ratios (inhale 3, hold 4, exhale 5).

Gradually extend as your capacity increases.

If "I feel lightheaded":

You're doing it correctly this can happen ini tially. If it's

uncomfortable, reduce the number of rounds or use shorter counts.

If "I fall asleep":

Perfect! This is an excellent practice before bed. If you need

to stay awake, do it while sitting upright instead

of lying down.

If "I can't tell if it's working":

Before starting, rate your anxiety 1 10. After finishing, rate

it again. You'll notice the difference in the number.

INTEGRATION

Use BEFORE: Difficult conversations, presenta tions, important decisions

Use AFTER: Stressful situations, triggering news, anxiety attacks

Combine WITH: Meditation (do 4 7 8 first, then meditate)

Best time: Morning (to start day calm), evening (to settle nervous system)

Especially helpful for: Anxious personalities, ra cing minds, insomnia.

◆ ◆ ◆

The Humming Reset

Humming creates vibration that resonates through your entire body, particularly affecting your vagus nerve—a key regulator of emotional state. Hum a simple tone for 30-60 seconds, focusing on feeling the vibration in your chest and head. This sonic vibration literally changes your physical frequency, which shifts your emotional frequency in turn.

Practice: The Humming Reset

Time Required: 1 2 minutes

Best For: Rapid nervous system recalibration, vagus nerve activation, tension release

Frequency: As needed, 2 3 times daily, especially when stuck or anxious

Equipment Needed: None

BASIC VERSION (1 2 minutes)

1. Sit or stand in a comfortable position with your spine straight
2. Close your eyes (optional but helpful)
3. Take three deep breaths to center yourself
4. Begin humming a simple, steady tone any pitch that feels natural
5. Feel the vibration in your chest, throat, and head as you hum
6. Continue humming for 30 60 seconds continuously
7. Pay attention to the vibration moving through your entire body
8. When you finish, sit quietly for 10 seconds and notice the shift

The vibration itself is doing the work you don't need to "do" anything except maintain the hum and feel it.

ENHANCED VERSION (3 5 minutes)

Add intentional frequency progression:

1. Start with a low hum for 20 seconds (feel it in your belly)
2. Gradually raise the pitch, moving the vibration upward through your body over 20 seconds
3. Peak at a higher pitch for 20 seconds (feel it in your head and crown area)
4. Gradually lower the pitch back down over 20

seconds

5. End with a low hum for 20 seconds (feel it in your belly again)

This journey through different frequencies creates a more complete nervous system reset and activates different energy centers.

VARIATION OM Humming:

Instead of a simple hum, chant "Om" (pronounced "Aum") repeatedly for 90 seconds. This ancient frequency is even more powerful for nervous system recalibration.

SUCCESS INDICATORS

Physical Signs: Vibration felt throughout body, particularly inchest and head; jaw relaxes; neck and shoulder tension releases;

teeth may feel tingly; head may feel clearer; nervous system visibly calms (you may notice your shoulders dropping)

Mental Signs: Mental chatter quiets; racing thoughts slow; mind becomes clearer and more fo cused; sense of inner peace increases;

anxiety diminishes

Energetic Signs: Blockages in energy channels begin to clear; sense of vibration moving through spine; energy becomes more unified and coherent; feeling of wholeness increases

Timeline: Immediate effect (vibration felt within

5 10 seconds),noticeable nervous system shift within 30 seconds, lasting effect

15 30 minutes

TROUBLESHOOTING

If "I don't feel the vibration":

You might be humming too softly. Hum louder so you can definitely feel it. The vibration should be palpable in your chest.

If "My throat gets tired":

You're either humming too loudly or too strained. Hum gently but audibly. Your throat should feel relaxed, not worked.

If "I can't hum at work/in public":

You can do a silent hum (just moving your lips without sound) or use the 4 7 8 Breath instead.

If "I don't notice any effect":

Try the Enhanced Version with the pitch pro gression. The journey through frequencies is more noticeable than a single pitch.

If "It feels awkward or silly":

This is completely normal. Humming activates the vagus nerve in profound ways the effect is real even if it feels odd.

INTEGRATION

Use BEFORE: Meditation or manifestation visualiza

tion (activates receptivity)

Use AFTER: Stressful conversations or emotional upheaval (releases trapped energy)

Combine WITH: The 4 7 8 Breath for deeper nervous system reset

Especially useful for: Anxiety, racing thoughts, muscle tension,energetic blockages, unclear think ing

Best timing: Morning (to start day calm), evening (to prepare for sleep)

Can be used: Anytime, anywhere (quietly if needed)

◆ ◆ ◆

The Nature Connection

Natural environments vibrate at frequencies that automatically harmonize human energy fields. When possible, step outside and make direct contact with nature—put your bare feet on grass, hug a tree, or simply gaze at the sky. Even two minutes of nature connection can significantly reset your vibrational state.

Practice: The Nature Connection

Time Required: 2 15 minutes (minimum 2 minutes for noticeable effect)

Best For: Grounding nervous system, releasing stuck emotions, full body frequency reset

Frequency: Daily ideal, minimum 3x weekly

Equipment Needed: Access to outdoors (even a single tree or patch of grass works)

BASIC VERSION (2 5 minutes)

Choose ONE of these nature connection methods:

1. **Barefoot Grounding**: Remove your shoes and socks. Stand with bare feet directly on grass, soil, or natural ground for 2 5 minutes. Feel the

earth's frequency entering through your feet.

2. **Tree Hugging**: Stand or sit with your back against a tree trunk. Place your hands on the tree. Close your eyes and feel the tree's life force and stability. 2 3 minutes minimum.

3. **Sky Gazing**: Stand outside and gaze upward at the sky for 2 5 minutes. If cloudy, that's fine. Your nervous system receives the light frequencies regardless. Take three deep breaths while looking at the sky.

4. **Water Touch**: If available, place your hands or feet in water (ocean, lake, stream, even a basin). Feel the flow and frequency of water for 2 5 minutes.

5. **Plant Contact**: Sit near plants or flowers. Touch their leaves gently. Feel their aliveness and vitality. Spend 2 5 minutes in their presence.

Natural environments vibrate at frequencies (around 7.83 Hz the Schumann Resonance) that automatically harmonize human energy fields.

ENHANCED VERSION (10 15 minutes)

Combine elements for more complete reset:

1. Start with barefoot grounding for 3 minutes (if possible)
2. Find a tree and place your back against it for 3 minutes

3. Gaze upward at the sky for 2 minutes

4. Sit quietly on the ground and feel the earth beneath you for 2 3 minutes

5. Notice any shifts in your energy, mood, or men tal clarity

This complete nature immersion creates profound nervous system recalibration and is excellent for releasing deep stress or emotional blockages.

SUCCESS INDICATORS

Physical Signs: Body temperature normalizes; breathing deepens; muscle tension releases; sense of heaviness/groundedness increases; sometimes tingling in feet when barefoot grounding; fatigue lifts.

Mental Signs: Mental chatter quiets; mind becomes clearer; sense of peace and calm; worry decreases; perspective shifts; creative ideas emerge

Energetic Signs: Sense of rootedness; blockages dissolve; energy becomes more stable and coherent; feeling of being "held" by the earth; connection to something larger than yourself increases

Timeline: Noticeable effect within 1 2 minutes for some, 5 minutes for others; cumulative benefit over days of regular practice; lasting effect 30 60 minutes

TROUBLESHOOTING

If "I don't have access to nature":

Even a potted plant in your room works. Sitting by a window looking outside works. Even looking at nature photos or videos has some effect.

If "I feel self conscious being barefoot outside":

Do this in your backyard or a private space. Or keep shoes on and place your hands on the ground instead.

If "I don't feel anything":

Try the Enhanced Version with multiple nature connections combined.Also, some people need 5 10 minutes before noticing effects.

If "The weather is bad":

Stand by a window and gaze outside. The frequen cies reach you through the glass. Or practice on your porch/balcony.

If "It feels like a waste of time":

This is one of the most powerful nervous system resets available.

Track your anxiety/stress before and after on a 1 10 scale. The difference will prove its value.

INTEGRATION

Use BEFORE: Difficult conversations or meetings (grounding creates stability)

Use AFTER: Stressful events or news (releases ac cumulated stress)Combine WITH: The Gratitude Burst (gratitude nature powerful reset)

Best timing: Early morning (sets tone for day) or evening (releases day's stress)Especially helpful for: Anxiety, overwhelm, mental fog, disconnection, stress accumulation

Can be used: Daily without limitation nature connection has no negative side effects

Science note: The earth's magnetic field frequency (7.83 Hz, called the Schumann Resonance) naturally synchronizes with healthy brainwave patterns. This is why "grounding" is so effective.

◆ ◆ ◆

The Heart-Hand Technique

Place your right hand over your heart and your left hand on your forehead. Close your eyes and breathe slowly, imagining love flowing between your hands and throughout your body. This position creates an energetic circuit that helps regulate both heart and brain rhythms, facilitating a return to coherence.

Practice: The Heart Hand Technique

Time Required: 2 3 minutes

Best For: Heart coherence activation, emotional regulation, creating wholeness, trauma release

Frequency: As needed, or daily for deep emotional work

Equipment Needed: None

BASIC VERSION (2 3 minutes)

1. Sit or stand in a comfortable position with good posture
2. Place your RIGHT hand directly over your heart (center of your chest)

3. Place your LEFT hand on your forehead between your eyebrows (third eye area)

4. Close your eyes

5. Take slow, deep breaths breathing fully and naturally

6. As you breathe, imagine LOVE flowing from the hand on yourforehead down to your heart, then flow ing back up from your heart to your head

7. Visualize this love flowing through your entire body down your arms, through your spine, into your legs

8. Continue for 2 3 minutes, maintaining the feel ing of love and wholeness

9. When complete, rest your hands and sit quietly for 30 seconds

This creates an energetic circuit between your heart (feeling center)and your head (thinking center), unifying emotional and rational intelligence.

ENHANCED VERSION (5 7 minutes)

Add progressive wholeness:

1. Start with Basic Version as described above (3 minutes)

2. Keeping hands in position, shift visualization: imagine love flowing OUT from your heart in all directions forward, backward, sides, up, down (1

minute)

3. Feel this love expanding outward, creating a sphere of coherence around your entire body (1 minute)

4. Then imagine this sphere expanding outward to encompass everyone and everything you care about (1 2 minutes)

5. Return attention to yourself: feel yourself as a center of love, coherence, and wholeness

This expanded practice moves from personal heart coherence to radiating coherence outward power ful for manifestation work.

VARIATION Different Hand Placement:

If the hand positions feel uncomfortable, try:

Both hands over heart

Right hand over heart, left hand on belly

Hands in prayer position at heart center

SUCCESS INDICATORS

Physical Signs: Heart rate may slow; breathing deepens naturally;chest feels warm or tingly; shoulders relax; jaw unclenches; hands may feel warm; overall body relaxes

Mental Signs: Racing thoughts quiet; mental focus narrows to present moment; sense of clarity in creases; worries fade into background; mind feels peaceful

Energetic Signs: Heart center "opens" or feels less contracted;sense of energy circuit between head and heart becomes tangible; feeling of wholeness and integration; emotional and rational self feel uni fied; sense of self compassion increases

Timeline: Effect begins within 30 60 seconds; deep ens over 2 3 minutes;lasting effect 30 60 minutes; cumulative benefits with repeated practice

TROUBLESHOOTING

If "I can't imagine love flowing":

Instead of imagining love, just feel the warmth of your hands.The physical contact creates the ef fect. Visualization enhances it but isn't required.

If "I don't feel anything":

You might have heart guarding a protective mechanism after emotional hurt. Try the Enhanced Version which expands gradually. Consistency over days helps.

If "My arms get tired":

Use lighter hand contact. Or place your hands in your lap and just focus on the energy connection between heart and head.

If "I feel emotional or want to cry":

This is perfect. The Heart Hand Technique often releases trapped emotions. Allow whatever emerges to flow. This is healing.

If "It feels awkward or self conscious":

This is a deeply powerful practice. Your re sistance may indicate old heart wounds. Consistent practice gently heals these.

INTEGRATION

Use BEFORE: Manifestation work (creates heart co herence for authentic desire alignment)

Use AFTER: Emotional challenges or conflicts (heals emotional wounds and unifies self)

Combine WITH: The Gratitude Burst for amplified heart opening

Combine WITH: The 4 7 8 Breath for deeper nervous system integration

Especially helpful for: Emotional healing, self love development,heart blockages, feeling discon nected from emotions or intuitioninternal conflict between head and heart

Best for: People with trauma history, chronic stress, or difficulty accessing heart centered feelings

Can be extended: Use as a meditation for 10 15 minutes for deep emotional work

Science note: The heart has its own neural network (40,000 neurons) that operates independently from the brain. This practice aligns

these two intelligence centers your "head" and your "heart."

◆ ◆ ◆

The key to making these reset tools effective is using them before your frequency drops too low. Don't wait until you're in full-blown anger or deep sadness to attempt a reset. The moment you notice your vibration beginning to lower—when irritation first appears, when worry begins to creep in—apply one of these techniques.

With practice, these tools become second nature—quick vibrational adjustments you make throughout your day to maintain an emotional frequency that supports rather than sabotages your manifestation efforts.

Remember: your emotional frequency isn't just a passive response to circumstances—it's a vibrational broadcast that you can actively manage. These reset tools give you practical ways to take control of that broadcast, ensuring that what you're sending out matches what you want to receive back.

Now it's time to put theory into practice with a structured 21-day experiment using the "miracle frequency" of 528 Hz. This practice is designed as a controlled personal study to help you determine how this frequency affects your manifestation abilities.

Whether you're a devoted believer in the Solfeggio system or a curious skeptic, this protocol will give you direct experience with how 528 Hz influences your vibrational state and manifestation results.

The 21-Day 528 Hz Protocol

What You'll Need:

- 528 Hz audio (available via QR code in the book or at www.vibrationbook.com/528hz)
- Headphones (preferred) or quality speakers

- Manifestation journal (template provided below)
- 20 minutes daily for 21 consecutive days
- A specific manifestation goal to work with throughout the experiment

Preparation:

- Select Your Manifestation Focus
- Choose one specific area for manifestation during this 21-day period. This could be:
- A particular financial goal
- A relationship quality you want to develop
- A creative project you're bringing to life
- A health outcome you're working toward
- The key is specificity—rather than "more abundance," choose "receiving an unexpected $500" or "finding a new client worth $1000."
- Baseline Assessment

Before beginning the practice, rate the following on a scale of 1-10 in your journal:

- Your emotional state regarding this manifestation (1=fearful/doubtful, 10=confident/joyful)
- The clarity of your vision for this manifestation (1=vague, 10=crystal clear)
- Your sense of alignment with this manifestation (1=feels forced, 10=feels natural)
- Any resistance you feel toward receiving this manifestation (1=strong resistance, 10=no resistance)

Set Up Your Space

Create a dedicated space for this practice where you can sit comfortably without distractions. Consider adding elements that enhance the sense of sacredness: a candle, a crystal, or an object that symbolizes your manifestation goal.

Daily Practice (20 minutes):

- → Centering (2 minutes)
- → Sit comfortably with your spine straight
- → Take three deep breaths, exhaling completely
- → Place your hands over your heart center

State your intention silently or aloud: "I open myself to the transformative frequency of 528 Hz to align with [your specific manifestation]." 528 Hz Immersion (15 minutes) Start the 528 Hz audio at a comfortable volume Close your eyes and bring your awareness to the center of your chest

As you listen, imagine the 528 Hz frequency as emerald green light filling your heart center Allow this light to expand throughout your body See it dissolving any blocks or resistance to your manifestationVisualize your manifestation as already complete, feeling the emotions of this realityIf your mind wanders, gently return focus to the sound and the green light Integration (3 minutes)

When the 15 minutes conclude, keep your eyes closed Place your hands on your heart again Take three deep breaths, imagining you're breathing in the reality of your completed manifestation Express gratitude for this manifestation as if it's already occurred Slowly open your eyes

Daily Journaling:

Immediately after each session, spend 3-5 minutes recording in your manifestation journal:

- ❖ Subjective Experience
- ❖ How did you feel physically during the session?
- ❖ What emotions arose?
- ❖ Did you notice any sensations in your body, particularly in your heart area?
- ❖ Any visual images or insights that came through?
- ❖ Daily Ratings (1-10 scale)
- ❖ Today's emotional state regarding your manifestation
- ❖ Clarity of vision

- Sense of alignment
- Level of resistance
- Synchronicities & Results
- Note any unexpected events related to your manifestation
- Record any progress, however small, toward your goal
- Document any shifts in your thoughts or behaviors related to this area

Weekly Integration (Days 7, 14, and 21):

- On these days, add a deeper review to your journal:
- Compare your ratings from the beginning of the week to the end
- Identify patterns in how the frequency affects you
- Note any significant shifts in your relationship to your manifestation
- Adjust your approach if needed based on what you're learning

Final Assessment (Day 22):

After completing the 21-day protocol, conduct a comprehensive review:

- Compare your baseline ratings with your final day ratings
- List all tangible results related to your manifestation goal
- Describe changes in your relationship to this manifestation
- Evaluate the overall impact of the 528 Hz frequency on your manifestation process
- Decide whether to continue using this frequency in your practice

Tips for Maximum Benefit:

Consistency is key: Try to practice at the same time each day to establish a strong energetic pattern.

Hydration matters: Drink water before and after your session, as proper hydration enhances energetic conductivity.

Enhance with intention: Hold or wear items that represent your manifestation goal during the session.

Extend the influence: Play 528 Hz softly in your environment throughout the day when possible.

Combine with action: Take at least one concrete step toward your manifestation goal each day, however small.

Troubleshooting:

- If you feel restless during sessions, try a walking meditation with the frequency playing through headphones.
- If you fall asleep consistently, try practicing earlier in the day or in a seated position rather than lying down.
- If you experience resistance to the practice, journal about it—resistance often points to blocks that need clearing.

This 21-day protocol gives you a structured way to explore how the 528 Hz frequency affects your manifestation abilities. By maintaining consistent practice and detailed documentation, you'll gather personal evidence about whether and how this frequency enhances your vibrational alignment with what you wish to create.

Whether you experience dramatic shifts or subtle changes, this experiment will deepen your understanding of how sound frequencies interact with your energy field and manifestation process.

Now it's time to put theory into practice with a comprehensive ritual designed to identify and resolve the vibrational misalignments that may be blocking your manifestations. This Desire Alignment Ritual combines writing, sound, and visualization to create coherence between your conscious desires and your subconscious vibrational patterns.

This isn't a quick fix but a deep realignment process. Set aside 60-90 minutes in a private space where you won't be disturbed. The insights and shifts that emerge may surprise you, and the resulting alignment can dramatically accelerate your manifestation results.

Preparation:

Gather these materials:

1. Journal and pen
2. Candle and matches/lighter
3. Glass of water
4. Small bowl of salt
5. Something that represents your desire (object, image, or symbol)

Optional: singing bowl, chimes, or bell

Optional: essential oil that resonates with your desire (citrus for abundance, rose for love, etc.)

1. Create sacred space by:
2. Clearing clutter from your environment
3. Turning off electronic notifications
4. Playing soft instrumental music if desired
5. Sitting comfortably with good posture

The Complete Desire Alignment Ritual:

Part 1: Clarification & Purification (15-20 minutes)

Begin by clarifying your desire and purifying your energy field:

- Opening Invocation (2 minutes)
- Light your candle
- Take three deep breaths

State aloud: "I now create perfect alignment between all aspects of my being for the manifestation of my highest good. All that is not in alignment is now revealed and harmonized."

Desire Clarification (5-7 minutes)

In your journal, answer these questions:

What specifically do I desire to manifest?

Why do I desire this? What will it bring to my life?

How will I feel when this desire is manifested?

What qualities or experiences does this manifestation represent to me?

Be as detailed and honest as possible. This isn't about creating the "perfect" desire statement but about clarifying what you truly want.

Energy Field Clearing (3-5 minutes)

Hold the bowl of salt in your hands

Close your eyes and imagine any conflicting energies around your desire being drawn into the salt

Speak aloud: "I release all energetic patterns that conflict with the clear manifestation of my desire."

When complete, set the bowl aside (you'll dispose of the salt after the ritual)

Essence Extraction (5 minutes)

In your journal, perform the essence extraction process:

Write: "If I had [your desire], what would it give me?"

Answer this question honestly

Then ask: "And if I had that, what would it give me?"

Continue this inquiry, going deeper with each question

Stop when you reach a core essence that doesn't lead to anything else

Circle this core essence—it's the fundamental frequency you're seeking to embody.

Part 2: Misalignment Identification (20-25 minutes)

Now identify specific vibrational misalignments that may be

blocking your manifestation:

Conscious-Subconscious Dialogue (10 minutes)

In your journal:

Draw a line down the middle of a page

Label the left side "Conscious Desire" and the right side "Subconscious Response"

On the left, write: "I desire [your specific manifestation]"

Then shift to your non-dominant hand and write whatever response emerges on the right side

Continue this dialogue, asking questions with your dominant hand and allowing answers to emerge with your non-dominant hand

Look for resistance, fears, or contradictions that appear

Misalignment Checklist (5 minutes)

Review the seven common misalignments from the previous section. For each one, rate on a scale of 1-10 how much this misalignment might be affecting your manifestation:

Surface Desire vs. Core Programming (1-10)

Stated Goals vs. Energetic Broadcast (1-10)

Desired Outcome vs. Feared Outcome (1-10)

Primary Desire vs. Secondary Gains (1-10)

Individual Desire vs. System Harmony (1-10)

Current Desire vs. Past Trauma (1-10)

Conscious Desire vs. Soul Blueprint (1-10)

Circle the 2-3 highest-rated misalignments for focused attention.

Body Wisdom Consultation (5-7 minutes)

Close your eyes and bring your full attention to your body

Visualize your desire as manifested and complete

Slowly scan your body from head to toe, noticing any areas of tension, contraction, or discomfort

For each area of tension, gently ask: "What are you trying to tell me about this desire?"

Record in your journal any insights that arise from this body consultation

Part 3: Vibrational Realignment (25-30 minutes)

Now use sound, visualization, and writing to create new alignment:

Sound Clearing (5 minutes)

If you have a singing bowl or bell, sound it while setting the intention to clear discordant frequencies

If you don't have these tools, use your voice to create a humming sound that feels resonant

Direct this sound to the areas of your body that showed tension in the previous exercise

Continue until you feel a shift or release in these areas

Alignment Statements (7-10 minutes)

For each of your top misalignments, create a specific alignment statement in your journal:

For Surface Desire vs. Core Programming:

"I now align my subconscious beliefs with my desire for [manifestation]. It is safe and beneficial for me to experience this."

For Stated Goals vs. Energetic Broadcast:

"I now embody the frequency of [your desire] in every cell of my being. My entire field broadcasts this reality."

For Desired Outcome vs. Feared Outcome:

"I release all fear related to [your desire]. I trust the perfect unfolding of this manifestation for my highest good."

For Primary Desire vs. Secondary Gains:

"I honor the ways my current situation has served me. I now

create new ways to meet these needs while allowing [your desire] to manifest."

For Individual Desire vs. System Harmony:

"I harmonize my desire for [manifestation] with all systems I'm part of. This manifestation serves not only me but contributes positively to the whole."

For Current Desire vs. Past Trauma:

"I gently heal the past experiences that created resistance to [your desire]. I am safe to receive this manifestation now."

For Conscious Desire vs. Soul Blueprint:

"I align my personal desire with my soul's highest path. I am open to receiving this manifestation in its most aligned form for my evolution."

Read each relevant statement aloud three times, feeling the resonance of alignment growing stronger with each repetition.

Vibrational Embodiment (5-7 minutes)

Stand up and shake out your body for 30 seconds

Place your hand over your heart

Close your eyes and connect with the core essence you identified earlier

Begin to physically embody this essence—move your body in ways that express this frequency

Allow sounds to emerge that resonate with this frequency

Continue until you feel fully immersed in the vibrational state of your desire

Integration Visualization (7-10 minutes)

Return to your seated position

Hold the object that represents your desire

Close your eyes and visualize a golden light at the crown of your head

See this light descending through your body, illuminating and harmonizing every cell

As the light moves, see it dissolving any remaining resistance or misalignment

When the light reaches your feet, see it creating a complete circuit of aligned energy

Visualize this aligned energy extending outward, connecting with the universal field

See your manifestation forming in this field, drawn to you by perfect resonance

Feel the emotion of receiving this manifestation as if it's already complete

Part 4: Anchoring & Completion (5-10 minutes)

Finalize the ritual by anchoring the new alignment and creating a bridge to daily practice:

Written Declaration (3-5 minutes)

In your journal, write a new, aligned declaration of your desire that integrates the insights from this ritual. Begin with:

"From a state of complete alignment, I now call forth [your desire]. All aspects of my being—conscious and subconscious, mind and body, personal and transpersonal—are in perfect harmony with this manifestation. I embody the frequency of [core essence] now, knowing that this naturally attracts the physical manifestation of [specific desire]."

Add any specific insights or commitments that emerged during the ritual.

Physical Anchoring (1-2 minutes)

Take a sip of water, imagining it sealing this new alignment in every cell

Touch the object representing your desire, infusing it with your aligned energy

If using essential oil, place a drop on your wrists or over your heart

Closing Gratitude (1-2 minutes)

Place both hands over your heart

Express gratitude for this alignment process

Thank your subconscious mind for its willingness to harmonize

Acknowledge that manifestation is now unfolding in divine timing

Extinguish the candle, symbolizing the completion of the ritual

Integration Practices:

To maintain and strengthen the alignment created in this ritual:

Daily Reinforcement

Read your aligned declaration each morning and evening

Spend 2-3 minutes embodying the core essence frequency

Notice and celebrate evidence of alignment throughout your day

Weekly Check-In

Once a week, review your journal entries from the ritual

Note any new insights or shifts that have occurred

Address any new misalignments that may have surfaced

Monthly Renewal

Perform a simplified version of this ritual monthly

Update your aligned declaration based on your evolving understanding

Deepen your embodiment of the core essence frequency

Tracking Template:

Create a simple tracking system in your journal with these columns:

Date

Alignment level (1-10)

Evidence of alignment

Manifestation progress

Insights or adjustments needed

Record entries at least weekly to document your journey of increasing alignment and manifestation.

This Desire Alignment Ritual addresses the root causes of manifestation blocks rather than just treating symptoms. By creating coherence between your conscious desires and subconscious patterns, you remove the interference that has been canceling your manifestation efforts.

Remember that alignment is not a one-time achievement but an ongoing process. As you grow and evolve, new layers of misalignment may surface for integration. This isn't a sign of failure but of deepening awareness and expanding capacity to manifest at higher levels.

The most powerful manifestations emerge not from perfect technique but from perfect alignment—when all aspects of your being are broadcasting the same frequency without contradiction or interference.

THE SCIENCE OF SOUND

CHAPTER 3: NADA YOGA & THE POWER OF OM

One sound can reprogram thousands of years of ancestral trauma.

One frequency can realign your entire nervous system in 15 minutes.

This is the secret yogis have protected for millennia.

In this chapter, you'll learn the specific sound that becomes your

most powerful manifestation tool—and how to use it to dissolve

blocks instantly.

Nada Brahma

"The universe is sound."

This simple yet profound statement forms the cornerstone of an ancient understanding known as Nada Brahma—a Sanskrit term that translates to "the universe is sound" or "sound is God." This concept isn't just poetic; it represents one of the most sophisticated frameworks for understanding reality ever developed.

In the yogic tradition, particularly within the Tantric and Vedantic schools, sound isn't just something we hear—it's the fundamental creative force of the cosmos. Before light, before form, before matter, there was vibration. Sound, in this understanding, precedes and gives rise to physical reality.

The ancient yogis identified two types of sound: ahata nada (struck sound) and anahata nada (unstruck sound). Ahata nada refers to ordinary sounds we hear with our ears—music, speech, the rustle of leaves. Anahata nada, however, refers to the primordial vibration that exists beyond physical hearing—the cosmic sound current that flows through all creation.

This distinction is crucial for understanding manifestation. While ordinary sounds affect us on a surface level, the unstruck sound operates at the causal level of reality—the realm where manifestation begins before taking physical form.

Modern physics echoes this ancient wisdom with string theory, which proposes that the fundamental building blocks of the universe are not particles but tiny vibrating strings. The different vibration patterns of these strings determine the properties of all matter and energy. In essence, physical reality emerges from vibrational patterns—exactly what the yogis have taught for thousands of years.

The implications for manifestation are profound. If the universe is fundamentally sound or vibration, then learning to work with sound gives you access to the creative process at its source. You're no longer trying to manipulate effects; you're engaging with cause.

In the Nada Brahma understanding, everything has its own unique sound signature—its own "song." This includes not just physical objects but also states of being, emotions, and experiences. Abundance has a particular vibrational signature. So does health. So does loving relationship. When you learn to attune yourself to these signatures through sound practices, you naturally draw these experiences toward you.

The yogis developed sophisticated practices to work with this sound-based reality. Through mantra recitation, they learned to create specific vibrational states in the body and mind. Through meditation on inner sounds (nada yoga proper), they learned to perceive and navigate the subtle realms of vibration that give rise to physical experience.

What makes this approach so powerful for manifestation is that it bypasses the thinking mind and works directly with vibrational reality. You don't have to believe in it for it to work, just as you don't have to believe in gravity to be affected by it. The principles operate based on the fundamental nature of reality, not on your opinions about them.

When you begin to experience yourself as a vibrational being in a vibrational universe—when you start to perceive the world through the lens of Nada Brahma—manifestation becomes less about forcing change and more about aligning with the vibrational patterns of what you wish to experience.

This shift in perspective changes everything. You stop seeing yourself as separate from what you want to manifest and begin to recognize that both you and your desires are expressions of the same vibrational continuum. The boundary between "you" and "what you want" begins to dissolve, replaced by an understanding of resonance and attunement.

The ancient wisdom of Nada Brahma offers a radical reframing of manifestation: What if, instead of trying to get what you want, you learned to become the vibrational equivalent of what you want? What if manifestation is less about attraction and more about resonance—about matching your personal vibration to the vibration of what you wish to experience?

This is the profound invitation of Nada Brahma—to recognize yourself as sound, to experience the world as sound, and to create through the conscious use of sound. It's an invitation that, once accepted, transforms not just how you manifest but how you experience reality itself.

Mantras as Frequency Keys

Imagine having access to specific sound combinations that,

when vocalized correctly, could unlock particular energies and states of consciousness. These sonic keys would allow you to tune your entire being to specific frequencies associated with what you want to manifest.

This isn't fantasy—it's the science and art of mantras, one of humanity's oldest and most sophisticated technologies for working with vibrational reality.

The word "mantra" comes from two Sanskrit roots: "man" meaning mind, and "tra" meaning tool or instrument. Literally, a mantra is a tool for the mind. But this definition barely scratches the surface of how mantras actually function.

At their most basic level, mantras are sound formulas designed to create specific vibrational effects in the human energy system. Unlike ordinary words, which primarily convey meaning, mantras are engineered for their sonic properties—the way they resonate in the body and energy field when properly intoned.

The science behind mantra effectiveness involves several key principles:

Entrainment

When you repeatedly vocalize a mantra, you create a dominant frequency in your system. Through the principle of entrainment—where rhythms in proximity tend to synchronize—your brainwaves, heart rhythm, and even cellular vibration begin to align with this frequency. This is why even simple mantras can create profound shifts in consciousness when repeated with focus.

Resonance

Each part of your body has natural resonant frequencies. Certain mantras are designed to resonate with specific energy centers or organs. For example, the bija (seed) mantra "LAM" resonates with the root chakra at the base of the spine, while "YAM" resonates with the heart center. When you chant these sounds, you create sympathetic vibration in these areas, activating and balancing their energy.

Neurological Patterning

The repetition of mantras creates new neural pathways, literally rewiring your brain over time. This is why traditional practice often involves repeating a mantra 108 times daily for 40 days—this consistent practice establishes the mantra's vibrational pattern as a dominant frequency in your nervous system.

Subtle Anatomy Activation

In yogic understanding, mantras don't just affect the physical body but also the subtle energy channels (nadis) and centers (chakras). Specific mantras are designed to clear blockages and stimulate flow in this subtle anatomy, which directly impacts your vibrational state and manifestation abilities.

What makes mantras particularly powerful for manifestation is their ability to bypass the thinking mind and work directly with vibrational reality. When you recite a mantra, you're not just affirming a concept—you're embodying a frequency.

Different mantras serve different manifestation purposes:

"Om Shreem Maha Lakshmiyei Namaha" vibrationally aligns you with abundance and prosperity

"Om Gam Ganapataye Namaha" helps remove obstacles to manifestation

"Soham" ("I am That") aligns you with your highest self and purpose

"Om Namah Shivaya" purifies your energy field of lower vibrations that block manifestation

The key to effective mantra practice isn't just mechanical repetition but conscious resonance. When you chant a mantra, bring your full awareness to the sound vibration—feel it resonating in your body, notice how it affects your energy, allow yourself to become the sound rather than just making the sound.

This is why traditional mantra initiation (diksha) was considered essential—a teacher who had mastered the mantra's vibration could transmit not just the correct pronunciation but the actual energetic signature of the mantra to the student. While such formal initiation is valuable, you can still receive significant benefits from mantra practice on your own, especially with widely-shared mantras like Om.

What distinguishes mantras from ordinary affirmations is their tested efficacy over thousands of years. These aren't random sounds—they're precisely engineered vibrational tools refined through generations of practice and observation. They carry the accumulated power of millions of people who have used them throughout history, creating a collective resonance that amplifies their effect.

When you work with mantras as part of your manifestation practice, you're not just changing your thoughts about what's possible—you're changing your vibrational frequency at a fundamental level, aligning yourself with the frequency of what you wish to manifest.

In a universe where everything is vibration, mantras give you direct access to the vibrational control panel of reality.

PRACTICE: The 108 Om Meditation - Foundation

Practice

Time Required: Beginner 10 12 min | Intermediate 20 25 min | Advanced 45 60 min

---

Best For: Attuning to primordial frequency, foundation for all manifestation

Frequency: Daily (ideal), minimum 3 4x weekly for results

Equipment Needed: None (optional: mala beads for counting, meditation cushion)

BEGINNER LEVEL: 27 Oms (10 12 minutes)

Perfect for establishing the practice and learning proper technique.

PREPARATION (2 3 minutes)

Physical Setup:

Sit comfortably with your spine straight but not rigid

If sitting on the floor, use a cushion to elevate your hips slightly above your knees

If in a chair, ensure both feet are flat on the ground

Place your hands in chin mudra: thumb and index finger lightly touching, other fingers extended

Close your eyes

Energetic Preparation:

Bring your awareness to the center of your chest

Imagine a golden sphere of light at your heart center

With each breath, see this light growing brighter and more radiant

Set your intention: "I align with the primordial frequency of creation for my highest manifestation"

THE THREE PART OM STRUCTURE

Each Om consists of three distinct sounds plus silence:

1. A (Ah) Begin from the base of your throat, mouth wide open
2. U (Ooo) Gradually close your mouth, sound moving to the middle throat
3. M (Mm) Close your lips completely, vibration moving to your head
4. SILENCE The unstruck sound that follows (most powerful part)

THE BEGINNER 27 OM PROTOCOL

Sound Technique:

Each Om should last 10 15 seconds total

Spend equal time on A U M (about 3 4 seconds each)

Allow 2 3 seconds of silence between each Om

Volume should be comfortable—not whispering, not shouting

Focus on feeling the vibration in your body more than hearing the sound

BREATH PATTERN:

Inhale fully before beginning each Om

Use your entire exhale for the Om sound

Never strain or force the breath

If you run out of breath, simply take another inhale and continue

Mental Focus (Three Phases):

PHASE 1 Oms 1 9 (Sound Awareness):

Focus on the sound itself

Feel it vibrate through your throat and chest

Don't try to change or control the sound—just experience it

PHASE 2 Oms 10 18 (Body Resonance):

Shift focus from the sound to the vibration

Feel the vibration spreading throughout your en tire body

Notice where in your body you feel the resonance most strongly

PHASE 3 Oms 19 27 (Energy Field Extension):

Focus on the vibration extending beyond your body

Feel it radiating into your energy field

Imagine the vibration creating ripples in the

space around you

After completing 27 Oms:

Sit in silence for 1 2 minutes

Notice any sensations, shifts, or insights

Feel the residual vibration settling into your system

INTERMEDIATE LEVEL: 54 Oms (20 25 minutes)

Once you can complete 27 Oms comfortably with good technique.

Enhanced Sound Technique:

Each Om extends to 15 20 seconds

Begin to experiment with different pitches—let your voice find its natural resonance

Add subtle variations in volume, getting slightly louder through the Om

Pay special attention to the silence after each Om—this is where the real power lies

Mental Focus (Three Phases):

PHASE 1   Oms 1 18 (Body Attunement):

Focus on how each Om attunes different systems of your body

Oms 1 6: Throat and voice (physical)

Oms 7 12: Chest and heart (emotional)

Oms 13 18: Head and mind (mental)

PHASE 2 Oms 19 36 (Chakra Activation):

Direct the vibration to specific energy centers

Oms 19 24: Root chakra (security, grounding)

Oms 25 30: Heart chakra (love, connection)

Oms 31 36: Third eye (vision, intuition)

PHASE 3 Oms 37 54 (Unified Field):

Let the focus expand to encompass your entire being

Feel all energy centers harmonizing

Sense the unified field of your complete self

Advanced Variations (Add after mastering basic 54 Oms):

MELODIC VARIATION:

Create a simple melody with each Om

Let the pitch rise gradually from Om 1 to Om 27

Then descend gradually from Om 28 to Om 54

This creates a more engaging practice and ampli fies the coherence effect

ENERGY CENTER FOCUS:

First 18 Oms: Root chakra (deeper voice, heavier vibration)

Next 18 Oms: Heart chakra (middle register, warm

vibration)

Last 18 Oms: Third eye (higher pitch, lighter vibration)

ADVANCED LEVEL: 108 Oms (45 60 minutes)

The traditional full practice for deep spiritual and vibrational transformation.

Mastery Techniques:

Each Om can extend 20 30 seconds as your breath capacity increases

Begin to hear the "unstruck sound" continuing even during silence periods

Practice becoming the vibration rather than just making the vibration

Allow spontaneous variations in pitch, volume, and duration as guided by your inner wisdom

Advanced Protocol Structure:

SECTION 1 GROUNDING (Oms 1 27): Root and Sacral Chakras

Deeper pitch, seated firmly

Focus: Stability, security, creative energy

Intention: "I ground my manifestations in the earth"

SECTION 2 ACTIVATION (Oms 28 54): Solar Plexus and Heart

Medium pitch, upright posture

Focus: Personal power, love, integration

Intention: "I activate my power to manifest from the heart"

SECTION 3 ELEVATION (Oms 55 81): Throat and Third Eye

Higher pitch, heart lifting

Focus: Expression, clarity, wisdom

Intention: "I see clearly and express my truth"

SECTION 4 TRANSCENDENCE (Oms 82 108): Crown Chakra

Pure vibration, minimal mechanical sound

Focus: Unity consciousness, divine connection

Intention: "I unite with the field of all creation"

Post Practice Integration:

Remain seated for 5 10 minutes in silence

Let any remaining vibration settle naturally

Journal any insights or experiences

Notice how your consciousness remains elevated

SUCCESS INDICATORS

Physical Signs:

Vibration felt intensely throughout entire body (especially chest, head, spine)

Throat and head may feel warm or tingly

Breathing becomes naturally slower and deeper

Body may sway slightly (don't resist this—it's the vibration moving through you)

Tingling sensations, especially in extremities

Overall relaxation and release of tension

Possible spontaneous movements or subtle body shifts

Mental Signs:

- Mental chatter significantly quiets or ceases entirely

Thoughts become less linear, more spacious

Sense of timelessness—the practice can feel much shorter than it actually was

Clarity increases; mental fog clears

Heightened creativity and insight

Some practitioners experience visual imagery or internal light

Energetic Signs:

Chakras feel activated and aligned (like spinning wheels)

Energy feels more coherent and integrated

Sense of wholeness, as though all fragmented parts unified

Heart area feels open and expanded

Connection to something larger than yourself

Energy field feels noticeably stronger and more stable

Sometimes sense of energy moving along the spine

Timeline:

Beginner (27 Oms): Effects noticeable within 5 minutes, lasting 30 60 minutes

- Intermediate (54 Oms): Deeper effects within 10 minutes, lasting 1 2 hours

Advanced (108 Oms): Profound effects by end, lasting several hours or longer

Cumulative Effects (First 40 Days of Daily Prac tice):

Days 1 10: Increasing coherence, deeper rest, better focus

Days 11 20: Noticeable manifestation acceler ation, intuitive clarity

Days 21 30: Significant nervous system recalibra tion, emotional stability

Days 31 40: Foundation fully integrated, Om be comes your default frequency

TROUBLESHOOTING

If "I'm self conscious about chanting out loud":

You can hum the Om instead of chanting it fully. The effect is almost identical.

Practice alone in your room with the door closed until you're comfortable.

Remember: authentic vibration matters more than volume.

If "I can't complete all the Oms without getting tired":

Start with fewer Oms (even just 3 9) and grad ually build up.

Focus on the quality of each Om rather than the quantity.

Your capacity will increase naturally with con sistent practice.

If "I don't feel anything happening":

You might be trying too hard or being too analytical.

Simply chant without expecting results—the vibra tion works regardless.

Try for 7 consecutive days before assessing the effect.

Some people feel subtle shifts they don't imme diately recognize.

If "My voice gets hoarse":

You're likely pushing too hard or straining.

Om should flow easily and naturally from your body.

If hoarse, reduce volume, take breaks between Oms, stay hydrated.

If "I feel emotional during or after practice":

This is completely normal. Om can bring up stored emotions.

Allow whatever emotions are present to flow.

This emotional release is part of the healing and clearing process.

If "I keep getting distracted by thoughts":

This is the entire point of the practice—to gently redirect attention.

Each time you notice distraction and return to the Om, you're strengthening your capacity.

Gradually, the mind naturally settles as it be comes entrained to the vibration.

If "Nothing seems to be manifesting faster":

27 Oms daily is the foundation. Consistency mat ters more than intensity.

Combine with the other practices in this book for cumulative effect.

Some results are internal (vibrational shifts) before they manifest externally.

INTEGRATION

Use BEFORE: Major manifestation rituals (attunes your system)

Use AFTER: Stressful events (resets your frequency to harmony)

Combine WITH: Binaural beats for enhanced brainwave entrainment

Combine WITH: Specific mantras for targeted mani festation work

Best timing: Early morning (sets frequency for en tire day)

Weekly: 1 2 times at 54 Oms for deeper work

Monthly: One full 108 Om practice for complete recalibration

DAILY PRACTICE PROGRESSION:

Week 1: 9 Oms daily (morning)

Week 2: 18 Oms daily (morning)

Weeks 3 4: 27 Oms daily (morning) Your foundation

is established

Months 2 3: 27 Oms daily 54 Oms on weekends

Month 4 : Can extend to full 108 Oms or maintain 27 54 daily

The 27 Om practice becomes your daily anchor. The more advanced versions deepen the work but are not necessary for significant manifestation results.

◆◆◆

Energy Center Attunement Variation

Once comfortable with basic Om practice, you can direct the vibration to specific chakras for targeted manifestation work:

Root Chakra Om (LAM-AUM): Combine LAM with Om for grounding and security manifestations

Sacral Chakra Om (VAM-AUM): Combine VAM with Om for creative and relationship manifestations

Solar Plexus Om (RAM-AUM): Combine RAM with Om for personal power manifestations

Heart Chakra Om (YAM-AUM): Combine YAM with Om for love-based manifestations

Throat Chakra Om (HAM-AUM): Combine HAM with Om for communication manifestations

Third Eye Om (AUM): Pure Om directed to the forehead for wisdom manifestations

Crown Chakra Om: Om as pure silence, experienced at the top of your head

Daily Integration Practices

Morning Activation: 3 Oms immediately upon waking to set your frequency for the day

Pre-Manifestation Tuning: 9 Oms before any manifestation visualization or writing practice

Transition Clearing: 1 Om when transitioning between activities, to clear the energy field

Evening Integration: 21 Oms before sleep to integrate the day's manifestation work

CHAPTER 4: MODERN SOUND SCIENCE

Brainwave Entrainment

What if you could shift your brain into optimal manifestation mode in just 15 minutes—using nothing but sound?

Not through meditation, not through effort, but through a scientifically proven phenomenon called entrainment.

This chapter reveals the exact sound frequencies that reprogram your brainwaves for effortless manifestation.

Let's explore the primary brainwave states and their relationship to manifestation:

Delta Waves (0.5-4 Hz): The Deep Unconscious

Delta waves dominate during deep, dreamless sleep. This is the realm of the unconscious mind and the body's healing processes. While not directly useful for conscious manifestation, delta states allow access to the deepest levels of your being where core beliefs and patterns reside. Some advanced manifestation techniques involve programming intentions during transitions into and out of delta states.

Theta Waves (4-8 Hz): The Gateway to the Subconscious

Theta waves occur during deep meditation, light sleep, and the hypnagogic state between wakefulness and sleep. This is where the conscious and subconscious minds meet—a powerful zone for manifestation work. In theta:

The critical, analytical mind relaxes its grip

The subconscious becomes more accessible

Visualization is more vivid and impactful

New beliefs can be more easily installed

Creativity and intuition flow freely

Many manifestation masters consider theta the optimal state for imprinting new desires into the subconscious mind.

Alpha Waves (8-13 Hz): The Bridge State

Alpha waves dominate when you're relaxed but alert—during light meditation, daydreaming, or when you're in "flow." This state creates a bridge between your conscious and subconscious mind, making it excellent for:

Setting intentions

Creative visualization

Affirmations

Receiving intuitive insights about your manifestation path

Alpha is the "manifestation preparation zone" where you can consciously design what you want to imprint more deeply during theta states.

Beta Waves (13-30 Hz): The Active Mind

Beta waves occur during normal waking consciousness, especially when you're engaged in active thinking, problem-solving, or focused attention. While necessary for implementing manifestation plans in the physical world, high beta states can actually inhibit manifestation by activating the critical mind and creating resistance. Learning to downshift from high beta to alpha and theta is essential for effective manifestation.

Gamma Waves (30-100 Hz): The Transcendent Mind

Gamma waves are associated with peak experiences, transcendent states, and moments of sudden insight. These high-frequency brainwaves create whole-brain synchronization and are found in advanced meditators. In gamma states:

Manifestation can occur almost instantaneously

The sense of separation between you and what you desire dissolves

You experience reality as highly malleable and responsive

While gamma states typically can't be maintained for long periods, even brief experiences can dramatically enhance your manifestation abilities.

The revolutionary discovery that makes all this practical is brainwave entrainment—the brain's natural tendency to synchronize with external rhythmic stimuli. When exposed to a consistent frequency through sound, light, or other means, your brainwaves will gradually shift to match that frequency.

This means you can deliberately induce specific brainwave states conducive to different aspects of manifestation simply by listening to properly designed audio.

For example:

Alpha entrainment (8-13 Hz) can help you shift from stressed thinking to relaxed awareness—ideal for setting clear intentions

Theta entrainment (4-8 Hz) can help you access the subconscious mind to reprogram limiting beliefs that block manifestation

Alpha-theta border entrainment (7-8 Hz) can create a perfect balance of conscious direction and subconscious receptivity

The science behind this is well-established. EEG studies clearly show that properly designed entrainment stimuli reliably shift brainwave patterns in predictable ways. This isn't pseudo-science—it's applied neuroscience that gives you direct access to the brain states most conducive to different aspects of manifestation.

What makes brainwave entrainment particularly valuable for manifestation work is that it doesn't require years of meditation practice to access powerful states of consciousness. While traditional meditation remains invaluable, entrainment provides a technological shortcut that can produce similar brainwave patterns in a fraction of the time.

By understanding the relationship between brainwave states and manifestation, and using entrainment tools to deliberately access optimal states, you gain a level of conscious control over your manifestation process that few people ever achieve.

Your brain is the physical interface between your consciousness and the vibrational field of reality. Learning to modulate its frequencies gives you direct access to the control panel of manifestation.

Binaural Beats & Isochronic Tones

When it comes to using sound technology for brainwave entrainment, two methods stand out for their effectiveness and accessibility: binaural beats and isochronic tones. These aren't just fancy names for meditation music—they're precisely engineered audio technologies designed to shift your brainwave patterns in specific, predictable ways.

Let's explore how each works and when to use them for maximum benefit in your manifestation practice:

Binaural Beats: The Stereo Brain Hack

Binaural beats work through a fascinating quirk of neurology. When you hear slightly different frequencies in each ear (for example, 200 Hz in your left ear and 210 Hz in your right), your brain creates a third "phantom" beat equal to the difference between the two tones (in this case, 10 Hz).

This phantom beat doesn't exist in the actual audio—it's generated by your brain as it tries to reconcile the difference between what your ears are hearing. This process, called "frequency following response," causes your brainwaves to synchronize with this phantom beat.

For example, if you want to induce a theta state (4-8 Hz) for deep manifestation work, you might listen to:

200 Hz in your left ear

204 Hz in your right ear

Your brain generates a 4 Hz binaural beat, gradually shifting you into theta

The key characteristics of binaural beats:

Require headphones: Since each ear must receive a different frequency, headphones are essential

Subtle effect: The entrainment happens below conscious awareness

Gentle transition: The shift in brainwave state occurs gradually

Works well with music: Can be embedded in pleasant background music

Effective for deep states: Particularly good for reaching theta and delta states

Isochronic Tones: The Direct Approach

Unlike the subtle approach of binaural beats, isochronic tones work through direct, rhythmic stimulation. They consist of a single tone that turns on and off rapidly at specific intervals, creating a distinct, evenly-spaced pulse.

For example, an 8 Hz isochronic tone (to induce alpha state) would simply turn on and off 8 times per second. This creates a strong, clear signal that the brain can easily follow.

The key characteristics of isochronic tones:

Don't require headphones: Work through regular speakers (though headphones enhance the effect)

More noticeable: You can consciously perceive the pulsing

Faster effect: Generally produce more rapid brainwave entrainment

Less pleasant sound: The pulsing can be more intrusive than binaural beats

Effective for alert states: Particularly good for alpha and beta entrainment

When to Use Each for Manifestation

Both technologies have their place in a comprehensive manifestation practice. Here's when to use each for optimal results:

Use Binaural Beats When:

Programming the subconscious: The subtle nature of binaural beats makes them ideal for deep work with the subconscious mind. Use them when implanting new beliefs or visualizing desired outcomes in detail.

Bedtime manifestation: Their gentle nature makes binaural beats perfect for pre-sleep manifestation work when you're transitioning into theta and delta states naturally.

Extended sessions: For longer manifestation sessions (30+ minutes), binaural beats are less fatiguing to the auditory system.

Emotional healing: When working with emotional blocks to manifestation, the gentle approach of binaural beats creates a safe container for processing difficult feelings.

Use Isochronic Tones When:

Energy activation: Their direct stimulation makes isochronic tones excellent for activating and directing energy toward your manifestation goals.

Focus enhancement: When you need to concentrate intensely on crafting clear intentions, isochronic tones in the alpha-beta border range (12-14 Hz) can create ideal focused awareness.

Quick state changes: When you need to rapidly shift your state before a manifestation practice, isochronic tones produce faster entrainment.

Group settings: Since they don't require headphones, isochronic tones can be used in group manifestation sessions.

Combining Technologies for Maximum Impact

Many advanced entrainment programs combine both technologies, using:

Isochronic tones to quickly establish the target brainwave state

Binaural beats to deepen and maintain that state

Carefully designed background music to enhance emotional resonance

This combination creates a powerful entrainment experience that addresses both the conscious and subconscious aspects of manifestation.

By understanding the strengths and applications of these sound technologies, you can strategically incorporate them into your manifestation practice, using the right tool at the right time to optimize your brainwave states for different manifestation tasks.

These aren't just pleasant sounds to relax to—they're precision instruments for accessing the brainwave states where manifestation happens most effectively.

Practice

Brainwave Entrainment for Manifestation Mastery

Your brain is a sophisticated biological computer that operates at different frequencies depending on your state of consciousness. By learning to deliberately shift these brainwave patterns using sound technology, you gain unprecedented control over your manifestation abilities.

Understanding Your Brainwave Manifestation Map

Delta (0.5 4 Hz): Deep unconscious programming and healing

- Best for: Overnight manifestation programming, deep belief restructuring
- Avoid for: Active visualization or conscious intention setting

Theta (4 8 Hz): The manifestation powerhouse

- Best for: Subconscious programming, vivid visualization, emotional healing
- Peak manifestation zone: 6 7 Hz where conscious and unconscious minds merge

Alpha (8 13 Hz): The bridge state

- Best for: Setting clear intentions, creative problem solving, receiving insights
- Optimal for beginners: 10 Hz provides stable,

accessible altered state

Beta (13 30 Hz): Implementation and action

- Best for: Planning manifestation steps, taking inspired action
- Avoid high beta (20 Hz) during visualization as it creates resistance

Gamma (30 100 Hz): Transcendent breakthroughs

- Best for: Moments of instant manifestation, spiritual connection
- Cannot be sustained long term but powerful for breakthrough moments

◆ ◆ ◆

Protocol 1: The Alpha Gateway (Beginner Practice)

Purpose: Establish reliable access to manifestation-conducive brainwave states
Duration: 15-20 minutes daily
Best Time: Morning or before dedicated manifestation work
Equipment: Headphones, 10 Hz binaural beat audio

The 15-Minute Alpha Gateway Protocol

Minutes 1-3: Descent Phase

- Focus on your breath without trying to control it
- Allow the binaural beats to begin shifting your brainwaves
- Notice physical sensations of relaxation beginning
- Don't judge or analyze—simply observe and allow

Minutes 4-10: Alpha Stabilization Phase

- You should now feel a distinct shift into relaxed alertness
- Your thoughts become less linear and more flowing
- This is optimal time for manifestation work

Minutes 11-15: Integration Phase

- Begin preparing to return to normal consciousness
- Take any insights or experiences and imagine filing them in your subconscious mind
- Express gratitude for the session
- Slowly become aware of your physical body

Protocol 2: Theta Reprogramming (Intermediate Practice)

Purpose: Deep subconscious restructuring and belief installation
Duration: 25-30 minutes, 3-4 times per week
Equipment: Headphones, 6 Hz binaural beat or theta isochronic tones

The 30-Minute Theta Reprogramming Protocol

Minutes 1-5: Induction Phase

- Begin with normal breathing, allowing the theta tones to gradually slow your brainwaves
- Use progressive muscle relaxation: tense and release each muscle group from feet to head

Minutes 6-10: Deepening Phase

- Allow your awareness to become diffuse and floating
- You may experience hypnagogic imagery—random pictures or scenes

- Don't try to maintain normal logical thinking—embrace the dream-like state

Minutes 11-25: Core Reprogramming Phase

Choose ONE focus for each session:

- *Belief Installation*: Replace limiting beliefs with empowering ones
- *Emotional Clearing*: Transform emotional blocks to manifestation
- *Future Memory Installation*: Create vivid "memories" of manifestation complete

Minutes 26-30: Integration and Emergence

- Stop active programming and simply rest in the theta state
- Begin slowly returning to normal awareness
- Count slowly from 1 to 10, becoming more alert with each number

CHAPTER 5: SOLFEGGIO & HEALING FREQUENCIES

Ancient Tones Modern Claims

One frequency (528 Hz) is called the "miracle frequency."

Thousands of people report life-changing results using it. Is it the frequency itself—or the alignment it creates?

In this chapter, you'll run your own experiment with 528 Hz and discover

whether this frequency becomes your personal manifestation accelerator. While their historical origins remain somewhat mysterious, their effects on human consciousness and manifestation abilities have been reported by thousands of practitioners worldwide.

At the core of the Solfeggio system are six primary frequencies, each associated with specific effects on body, mind, and spirit:

396 Hz: Liberation from Fear and Guilt

This frequency resonates at the root of many manifestation blocks. It's associated with releasing guilt, fear, and defensive patterns that create energetic resistance to what you're trying to manifest. When working with this frequency:

Emotional debris from the past begins to dissolve

Security-based fears that block abundance lose their grip

The energy of your root chakra stabilizes, creating a solid foundation for manifestation

---

Practice: 396 Hz Fear & Guilt Release

Time Required: 15 20 Minutes Daily

Best For: Clearing root chakra fears blocking abun dance manifestation, releasing guilt

Frequency: Daily for 21 consecutive days minimum

Equipment Needed: 396 Hz audio,
headphones or speakers, journal

BASIC VERSION (15 minutes)

PREPARATION (2 minutes)

Find a safe, quiet space where you feel secure

Sit or lie down comfortably

Place your hand over your lower abdomen (root chakra area)

State intention: "I now release all fear and guilt that blocks my abundance"

LISTENING & RECEIVING (10 minutes)

Start 396 Hz audio at comfortable volume

Close your eyes

Visualize deep red light at the base of your spine

Feel this light beginning to vibrate and release

With each breath, imagine fear and guilt being drawn out like smoke

Don't try to force anything—simply allow

Notice any emotions, sensations, or memories that surface

Allow them to move through without resisting

INTEGRATION (3 minutes)

Let the audio continue softly

Place both hands on your heart

Feel yourself as fundamentally safe

State: "I am safe. I deserve abundance. I am free from guilt."

Take three deep breaths

Open eyes slowly

ENHANCED VERSION (20 minutes)

PREPARATION (2 minutes)

As in basic version

GROUNDING PHASE (3 minutes)

Before starting audio, ground yourself

Feel your connection to the earth

Imagine roots growing from your feet into the earth

Feel support rising up from the earth into your body

ROOT CHAKRA ACTIVATION (2 minutes)

Visualize red light at base of spine

Gently contract and release pelvic floor muscles 5 10 times

This activates the root chakra and readies it for release

DEEP LISTENING & RELEASE (10 minutes)

Start 396 Hz audio

Close eyes

Bring specific fears to mind one at a time

For each fear:

Acknowledge it: "I see this fear"

Feel it: Notice where in your body it lives

Release it: Imagine it dissolving in the 396 Hz vibration

Work through 3 5 primary fears or guilt patterns

EMBODIMENT (3 minutes)

Visualize yourself free from fear

Feel your body relaxing more deeply

Breathe naturally and freely

State: "Fear and guilt are releasing from my

being now"

SUCCESS INDICATORS

Physical Signs: Tingling at base of spine, warmth spreading through lower body,

spontaneous stretching or movement, relaxation of pelvic floor muscles,

deepening of breath, possible trembling (this is release)

Mental Signs: Clarity about guilt sources, refram ing of fearful situations,

diminished intensity of worry, insights about past traumas,

noticeable shift from defensive thinking to open ness

Emotional Signs: Release of heaviness, tears (es pecially valuable),

increase in confidence, reduction in worry spirals,

feeling of genuine safety, possible spontaneous forgiveness

Timeline:

First session—noticeable shift within 5 10 minutes

Week 1—Measurable reduction in baseline anxiety

Week 3—Significant emotional clearing, new per spectives on fears

Weeks 4 —Integration into new baseline, persistent ease

Troubleshooting

If "I feel emotional or tearful":

This is perfect. Tears are how the body releases stored fear and guilt.

Allow the full emotional release.

If "I don't feel anything":

You might be trying too hard. Simply listen without expectations.

Subtle shifts count—notice if you're slightly less worried afterward.

If "Intense emotions or memories surface":

Continue the practice. This is the fear being released.

Pause if overwhelming, resume when ready.

If "Physical sensations feel uncomfortable":

This is usually release, not harm. Trembling or tingling is normal.

If truly uncomfortable, take a break and resume later.

INTEGRATION

Use BEFORE: Financial decisions or abundance mani

festation work

Use AFTER: Stressful situations that trigger fear

Combine WITH: The Gratitude Burst for deepening safety feeling

Best time: Evening, allowing subconscious processing overnight

Progression: Use daily for 21 days, then 3x weekly

◆◆◆

417 Hz: Transformation and Change

This frequency facilitates breaking out of stagnant situations and old patterns. It resonates with the sacral chakra, associated with creativity and emotional fluidity. When working with 417 Hz:

- Rigid thought patterns become more flexible
- Resistance to necessary change diminishes

Creative solutions to manifestation challenges emerge more readily

---

Practice: 417 Hz Breaking Patterns & Facilitating Change

Time Required: 15 20 minutes daily

Best For: Breaking stuck patterns, clearing resistance to necessary change

Frequency: Daily during periods of desired transformation (14 28 days)

Equipment Needed: 417 Hz audio, headphones, journal, optional: movement space

BASIC VERSION (15 minutes)

PREPARATION (2 minutes)

Identify the stuck pattern you want to break

Sit or stand in an open posture

State intention: "I now break free from [specific pattern]"

Allow yourself to feel your readiness for change

LISTENING & ENERGETIC SHIFTING (10 minutes)

Start 417 Hz audio

Close eyes

Visualize orange light filling your sacral chakra area

See stagnant patterns as grey/brown smoke being dissolved

Imagine new energy flowing into the space created by this release

With each breath cycle:

Inhale: Notice new possibility

Exhale: Release the old pattern

Let the 417 Hz frequency do the work of dissolving rigidity

INTEGRATION & NEW DIRECTION (3 minutes)

Visualize yourself living free from this pattern

Feel the difference in your body and energy

State: "I am transformed. New

possibilities flow through me now."

Express gratitude for this release

ENHANCED VERSION (20 minutes ACTIVE TRANSFORMATION)

PREPARATION (2 minutes)

As in basic version

PATTERN IDENTIFICATION (2 minutes)

Journal quickly: What pattern am I breaking?

What evidence shows this pattern is stuck?

What will be possible when I'm free from this?

MOVEMENT & RELEASE (5 minutes)

Start 417 Hz audio

Stand with arms at sides

Begin gentle swaying with the frequency

Gradually increase movement:

Minute 1 2: Subtle swaying

Minute 3: Add arm movements

Minute 4 5: Full body movement expressing release

Move as if physically shaking off the old pattern

Allow sounds to emerge

STILLNESS & REPROGRAMMING (8 minutes)

Sit or lie down

Continue 417 Hz audio

Breathe deeply

Visualize new pattern taking root:

See yourself behaving differently

Feel the emotions of this new pattern

Sense the physical ease of this transformation

Imagine this pattern fully integrated into your life

FINAL ANCHORING (3 minutes)

State transformation as complete: "I am free from [old pattern]"

Feel the 417 Hz frequency anchoring this change into your being

Express commitment to the new pattern

SUCCESS INDICATORS

Physical Signs: Tingling sensation flowing through body, spontaneous stretching,

release of held tension, feeling of movement or flow,

possible yawning or deep sighing, energy activation

Mental Signs: Sudden insights about the pattern, new perspectives emerging, mental flexibility increases, problem solving improves, resistance to change noticeably decreases

Emotional Signs: Excitement about possibilities, relief at prospect of change,

decreased rigidity in thinking, openness to new ways,

feeling of aliveness and renewed potential

Behavioral Signs: Natural shifts in habitual behaviors, easier decision making about change, increased motivation for new actions, spontaneous words or phrases indicating transformation

Timeline: First session—noticeable mental shift, increased openness

Days 1 7—Initial behavioral

changes emerge naturally

Days 8 14–New pattern feeling more natural and sustainable

Weeks 3 4–Complete integration of new pattern

TROUBLESHOOTING

If "I feel resistance or don't want to change":

This reveals secondary gains from the pattern.

Explore what the old pattern gives you.

Use 417 Hz specifically for this resistance.

If "I understand intellectually but don't feel the shift":

Use the Enhanced Version with movement.

Physical movement enhances the frequency's effect.

If "Change feels scary":

Completely normal. Use 396 Hz first to address underlying fears,

then return to 417 Hz once safe feeling is established.

If "I keep returning to the old pattern":

One session isn't enough for deeply rooted patterns.

Use 417 Hz daily for minimum 21 days.

INTEGRATION

Use BEFORE: Major life transitions, deliberate behavior changes

Use AFTER: Recognizing you're stuck in old patterns

Combine WITH: The Physical Pattern Interrupt for embodied change

Best time: Morning (sets new pattern for the day)

Progression: Daily during transformation, then 2x weekly for maintenance

◆ ◆ ◆

528 Hz: Transformation and Miracles (The "Love Frequency")

Perhaps the most celebrated of the Solfeggio tones, 528 Hz is said to resonate with the heart chakra and has been associated with DNA repair, transformation, and miraculous manifestations. When working with this frequency:

The heart field comes into greater coherence Self-love increases, allowing greater receptivity to manifestation. The gap between current reality and desired manifestation seems to narrow

---

Practice: 528 Hz The Love Frequency & Heart Activation

Time Required: 15 25 minutes daily

Best For: Heart centered manifestation, healing relationship wounds, self love development

Frequency: Daily for 33 consecutive days (sacred cycle)

Equipment Needed: 528 Hz audio, headphones or speakers, optional: heart shaped crystal

BASIC VERSION (15 minutes)

PREPARATION (2 minutes)

Sit comfortably with hand over heart

Take three deep breaths into your heart center

State intention: "My heart opens to give and receive love freely"

Feel yourself becoming receptive to love's frequency

LISTENING & HEART OPENING (10 minutes)

Start 528 Hz audio at warm volume

Close eyes

Visualize emerald or pink light at your heart center

Imagine this light growing with each heartbeat

Allow the frequency to gently open any heart guards or walls

Feel your heart softening, expanding, becoming more spacious

Notice the love that's already present within you

Don't try to change anything—simply allow opening

EMBODIMENT (3 minutes)

Visualize your heart light expanding outward:

First filling your entire body

Then extending 3 feet in all directions

Then radiating throughout your entire environ ment

Feel yourself as a transmitter of love

State: "I am love. I radiate love. I receive love."

ENHANCED VERSION (20 25 minutes RELATIONSHIP HEALING)

PREPARATION (2 minutes)

As in basic version

Optional: Hold a heart shaped crystal over your heart

HEART COHERENCE ACTIVATION (3 minutes)

Begin 528 Hz audio

Bring to mind someone you love easily (pet, child, dear friend)

Feel genuine love for them

Allow this love feeling to settle into your heart

Continue breathing this love

PERSONAL SELF LOVE PHASE (5 minutes)

Shift focus to yourself

Bring to mind moments when you felt truly loved and accepted

Allow yourself to receive that love fully

Place your hands on your heart and send yourself the same love

Say internally: "I love you [your name]. I accept you completely."

Feel this love filling any places in your heart where shame or doubt lived

SPECIFIC RELATIONSHIP HEALING (7 minutes)

Bring to mind any relationship that needs healing

See this person with love and compassion

Release blame or hurt: "I release this hurt with love"

Visualize sending healing light to this person

Ask your heart: "What did this relationship teach me?"

Send gratitude for the lessons

Visualize complete healing and peaceful closure (regardless of current status)

UNIVERSAL LOVE EXPANSION (5 minutes)

Expand your love outward to everyone and everything

See your heart light connecting with all beings

Feel the unity of all consciousness

Rest in this state of universal love

SUCCESS INDICATORS

Physical Signs: Warmth in chest, softening of fa cial muscles, relaxation of jaw,

heart feeling expansive or light, spontaneous smiling,

breathing becoming deeper and more natural, pos sible tears

Mental Signs: Shift from judgment to compassion, increased forgiveness capacity,

more positive thoughts about self and others,

mental chatter decreasing, sense of peace increas

ing

Emotional Signs: Genuine love and warmth emerging, sense of being held and safe,

reduction in loneliness, increased connection to others,

feeling of belonging and acceptance, joy arising

Relationship Shifts: Others responding with more warmth and openness,

increased affection and intimacy, conflict resolving more easily,

feeling of genuine connection deepening

Timeline: First session—immediate opening and softening

Days 1 7—Noticeable warmth in chest, increased smiling

Days 8 21—Relationships becoming warmer, self judgment decreasing

Days 22 33—Complete heart reset, compassion becoming default state

TROUBLESHOOTING

If "I don't feel love emerging":

You may have thick heart walls from past hurt. Be patient.

Continue practice daily; walls soften gradually.

If "I feel sad or emotional":

Perfect—buried grief is being released. Let it

flow.

This clearing is necessary for genuine love to emerge.

If "This feels indulgent or self centered":

Self love isn't selfish; it's the foundation for genuine giving.

The world needs your love filled presence.

If "Relationships don't change":

Your frequency is changing; relationships respond in time.

Some may not, and that's information you need.

INTEGRATION

Use BEFORE: Difficult conversations, relationship decisions, self acceptance work

Use AFTER: Heartbreak, betrayal, rejection

Combine WITH: Morning Frequency Routine for daily heart opening

Best time: Evening (allows heart opening during night)

Special: Complete 33 day cycle in full for maximum transformation

◆ ◆ ◆

639 Hz: Connection and Relationships

This frequency promotes harmonious interpersonal connections and balance between different aspects of self. It resonates with the throat chakra, associated with communication. When working with 639 Hz:

1. Communication blocks that hinder manifestation clear
2. Relationships that support your manifestation goals strengthen
3. Internal conflicts about what you're manifesting resolve

Practice: 639 Hz Relationship Harmony & Connection

Time Required: 15 20 Minutes

Best For: Improving communication, deepening relationships, resolving conflict

Frequency: 3 5 times weekly as ongoing practice, or daily during relationship challenges

Equipment Needed: 639 Hz audio, journal

BASIC VERSION (15 minutes)

PREPARATION (2 minutes)

Bring to mind a specific relationship

you want to harmonize

Sit in open posture (open heart, relaxed shoulders)

State intention: "I now harmonize my relationship with [person]"

LISTENING & ATTUNEMENT (10 minutes)

Start 639 Hz audio

Visualize blue green light in your throat chakra

Imagine this light extending to connect with the other person

See a beam of light connecting your hearts

Feel genuine understanding and compassion flowing both directions

Release any resentment or judgment

Visualize smooth, honest, warm communication between you

INTEGRATION (3 minutes)

Visualize the relationship healed and harmonious

Feel the ease of authentic connection

State: "We communicate with love and understanding"

ENHANCED VERSION (20 minutes FOR RESOLVING CONFLICT)

PREPARATION (2 minutes)

As in basic version

COMMUNICATION BLOCK IDENTIFICATION (3 minutes)

Journal: What's creating distance or conflict?

What needs to be said but hasn't been?

What is the other person trying to express?

THROAT CHAKRA ACTIVATION (2 minutes)

Sing or hum while listening to 639 Hz

Allow sounds to emerge naturally

Feel the throat chakra coming alive

AUTHENTIC EXPRESSION PHASE (6 minutes)

Imagine the other person sitting before you

Speak what needs to be said:

\\\"I need...\\\"

\\\"I feel...\\\"

\\\"I appreciate...\\\"

\\\"I forgive...\\\"

Speak as if they can truly hear you

Then listen to their response in your imagination

Notice what they might have been wanting to say

HARMONIZATION (5 minutes)

Visualize perfect understanding between you

See new communication patterns emerging

Feel mutual respect and genuine listening
Imagine the relationship naturally harmonizing

SUCCESS INDICATORS

Physical Signs: Relaxation in throat and jaw, voice becoming clearer,
ability to breathe more deeply, shoulders relaxing,
ease of expression, natural communication flow

Mental Signs: Clarity about relationship dynamics, new understanding of the other person,
less reactivity in thinking about them,
ability to see their perspective more easily

Emotional Signs: Reduced defensiveness, increased empathy for the other person,
less resentment, feeling of genuine connection,
increased warmth and affection

Relationship Changes: Others becoming more responsive and open,
conversations becoming easier and warmer,
conflict resolving faster, deeper understanding,
more authentic connection

Timeline: First session–noticeable ease in communication

Days 1 7–Relationship feeling warmer, conversations improving

Weeks 2 3–Conflict patterns dissolving, understanding deepening

Ongoing–Harmonious communication becoming the new normal

TROUBLESHOOTING

If "The other person doesn't change":

You can only change your frequency. Their response may come later,

or this relationship may need different boundaries.

If "I feel resistant to harmonizing":

There may be valid reasons to create distance.

Allow your intuition to guide boundaries.

If "I still have anger toward them":

Use 396 Hz first to process the anger,

then return to 639 Hz for harmony.

INTEGRATION

Use BEFORE: Difficult conversations, relationship decisions

Use AFTER: Conflict or disconnection

Combine WITH: Heart Hand Technique for embodied harmony

Group setting: Excellent for family harmony work

◆ ◆ ◆

741 Hz: Expression and Solutions

This frequency stimulates creative expression, intuitive insight, and problem-solving abilities. It resonates with the third eye chakra, associated with vision and intuition. When working with 741 Hz:

Creative blocks dissolve

Intuitive guidance about manifestation becomes clearer

Solutions to implementation challenges appear more readily

Practice: 741 Hz Creative Expression & Intuitive Solutions

Time Required: 15 20 Minutes

Best For: Clearing creative blocks, activating in tuition, solving stuck problems

Frequency: 2 3 times weekly, or daily when facing creative challenges

Equipment Needed: 741 Hz audio, journal, optional: artistic supplies

BASIC VERSION (15 minutes)

PREPARATION (2 minutes)

Identify the creative block or problem you want solved

Sit comfortably in upright posture

State intention: "My creative solutions now flow freely"

Open yourself to receiving insight

LISTENING & THIRD EYE ACTIVATION (10 minutes)

Start 741 Hz audio

Close eyes, bring awareness to space between eyebrows

Visualize indigo or purple light illuminating your third eye

Release the conscious effort to solve the problem

Enter a state of receptive listening

Allow images, ideas, or insights to flow without judgment

Notice what emerges naturally

CREATIVE EXPRESSION PHASE (3 minutes)

When ready, open eyes (keep audio playing if desired)

Quickly write or sketch what emerged

Don't edit or judge—simply capture the flow

Allow creative expression to continue naturally

ENHANCED VERSION (20 minutes DEEP CREATIVE ACTIVATION)

PREPARATION (2 minutes)

As in basic version

CREATIVE BLOCK EXPLORATION (3 minutes)

Journal: What am I not expressing?

What creative desires have I suppressed?

What fear blocks my creative flow?

INTUITIVE CHANNEL OPENING (5 minutes)

Start 741 Hz audio

Visualize indigo light at your third eye

Imagine this light opening like a door

See divine inspiration, creativity, and wisdom flowing through

Feel your intuition waking up and becoming active

Allow any creative impulse to emerge

CREATIVE EXPRESSION (8 minutes)

Let your hands move creatively:

- Write freely without stopping
- Draw or paint without planning
- Move expressively to the frequency
- Sing or sound intuitively

Allow the 741 Hz to guide your expression

Don't censor or judge the output

INTEGRATION (2 minutes)

Allow the creative expression to complete naturally

Journal: What did I learn from this?

What solutions or insights emerged?

SUCCESS INDICATORS

Physical Signs: Tingling between eyebrows, vision becoming sharper,

energy activation, hands wanting to move creatively,

voice wanting to express, body feeling activated

Mental Signs: Sudden ideas and insights, mental clarity increasing,

creative solutions appearing seemingly from nowhere,

problem solving becoming easier, intuition becoming clearer

Creative Shifts: Creative blocks dissolving, ideas flowing more easily,

creative output improving in quality and quantity,

new perspectives on stuck problems,

ability to express authentically

Intuitive Signs: Increased hunches and gut feelings that prove accurate,

better timing in decisions, synchronicities increasing,

connection to higher guidance strengthening

Timeline: First session—often immediate ideas or insights

Days 1 7—Creative flow noticeably easier

Weeks 2 4—Problems that seemed stuck suddenly have solutions

Ongoing—Creative expression becoming natural and abundant

TROUBLESHOOTING

If "Nothing comes to me":

Don't force it. Receptivity requires relaxation. Try again when you're less goal oriented.

If "Ideas come but feel mediocre":

First ideas are often just clearing the blocks. Continue practice; better ideas will follow.

If "My creative output doesn't match my vision":

That's the creative journey. Trust the process. Refinement comes with continued expression.

INTEGRATION

Use BEFORE: Creative work, brainstorming, problem solving sessions

Use AFTER: Receiving new insight (helps integrate it)

Combine WITH: Morning Frequency Routine to start day creatively

Group setting: Excellent for team creativity and innovation

◆◆◆

852 Hz: Spiritual Insight and Awakening

This highest of the core Solfeggio frequencies is associated with spiritual connection and the crown chakra. When working with 852 Hz:

Connection to higher guidance strengthens

Spiritual perspective on manifestation expands

Alignment with your highest purpose clarifies

Practice: 852 Hz Spiritual Insight & Divine Connection

Time Required: 20 30 Minutes

Best For: Spiritual alignment, higher guidance con nection, soul level manifestation

Frequency: 1 3 times weekly, especially during major life transitions

Equipment Needed: 852 Hz audio, quiet sacred space, journal

BASIC VERSION (20 minutes)

PREPARATION (2 minutes)

Create sacred atmosphere (candle, incense, essen tial oil optional)

Sit in meditation posture

State intention: "I now connect with divine wis dom for my highest path"

Acknowledge the intelligence greater than your self

LISTENING & CROWN ACTIVATION (15 minutes)

Start 852 Hz audio

Visualize violet or white light at crown of head

Feel this light connecting you to infinite intelligence

Release all personal concerns and desires

Open to seeing your life from a higher perspective

Ask: \\\"What do I need to know about my purpose? \\\"

Ask: \\\"How can I align my manifestations with my soul's path?\\\"

Listen for insights without forcing

Allow yourself to experience connection to some thing greater

INTEGRATION & GROUNDING (3 minutes)

Slowly bring awareness back to your body

Feel the divine connection integrated into your being

Journal any insights that came through

ENHANCED VERSION (30 minutes COMPLETE SPIRITUAL

REALIGNMENT)

PREPARATION (2 minutes)

Create sacred space intentionally

Ground yourself with feet on earth or hands on ground

State comprehensive intention:

\\\"I align completely with divine wisdom for my highest path and purpose\\\"

GROUNDING TO EARTH (3 minutes)

Visualize roots extending from base of spine to earth center

Feel stability and earthly support

Acknowledge your human existence

CHAKRA ASCENSION JOURNEY (12 minutes)

Start 852 Hz audio

Visualize ascending through each energy center:

Red at root: Feel security and physical presence

Orange at sacral: Feel creative life force

Yellow at solar plexus: Feel personal power

Green at heart: Feel love and connection

Blue at throat: Feel authentic expression

Indigo at third eye: Feel expanded vision

Violet at crown: Feel divine connection

Feel each center lighting up and harmonizing

DIVINE COMMUNION (10 minutes)

Rest at crown connection

Ask the highest wisdom:

\\\"What is my soul's purpose?\\\"

\\\"How am I being called to evolve?\\\"

\\\"What do I need to align with my highest good?\\\"

\\\"How can my personal desires serve the greater whole?\\\"

Listen without expectation

Receive whatever comes through

Feel complete alignment with divine intelligence

INTEGRATION & GROUNDING (3 minutes)

Slowly return to body awareness

Feel divine wisdom integrated into every cell

Journal all insights

SUCCESS INDICATORS

Physical Signs: Tingling at crown, feeling of peace spreading through body,

expanded sense of space, sense of being held,

relaxation deepening, possible spontaneous energy movement

Spiritual Signs: Sense of connection to something greater,feelings of unconditional love, peace be yond normal understanding, sense of being guided, synchronicities aligning,clear knowing about deci sions

Intuitive Signs: Reception of clear guidance, ex

panded perspective on life,understanding of soul purpose clarifying,intuition becoming very clear and reliable,sense of higher intelligence directing flow.

Life Direction: Suddenly seeing your path clearly, decisions becoming obviously right or wrong,manifestations aligning with higher purpose,sense of being in perfect flow with life

Timeline: First session—often profound spiritual experience

Days 1 7—Sense of divine guidance becoming stronger

Weeks 2 4—Life direction clarifying significantly

Ongoing—Living from alignment with higher purpose

TROUBLESHOOTING

If "I don't experience spiritual connection":

Spiritual experience can't be forced. Simply show up.

Trust that the connection is happening even if not felt.

If "I receive conflicting guidance":

Not all impressions are divine guidance.

Learn to recognize the quality of true spiritual wisdom

(clarity, peace, alignment with highest good).

If "My personal desires conflict with spiritual guidance":

This is important information. Journal about the conflict.

True spiritual path usually integrates personal fulfillment.

INTEGRATION

Use BEFORE: Major life decisions, career changes, relationship commitments

Use AFTER: Confusion or sense of being lost

Combine WITH: 108 Om meditation for deepest spir itual work

Special: Monthly or quarterly for complete perspec tive realignment

◆ ◆ ◆

What makes these frequencies particularly interesting for manifestation work is their reported ability to clear blockages at different levels of your being. Many manifestation efforts fail not because the technique is wrong but because unresolved energetic patterns create interference.

The Solfeggio frequencies appear to work directly with these energetic patterns, creating resonance that helps dissolve disharmony and establish coherence. This creates a cleaner vibrational field from which to broadcast your manifestation intentions.

The rediscovery and popularization of these frequencies in modern times is largely attributed to Dr. Joseph Puleo, who claimed to have found them encoded in the Bible's Book of Numbers. While this specific origin story remains controversial, the frequencies themselves have gained widespread use in sound healing communities.

Modern sound healers often combine these pure tones with music, creating compositions that are both aesthetically pleasing and therapeutically effective. Some practitioners also use tuning forks tuned to these frequencies for more targeted application to specific body areas or energy centers.

What's particularly notable about the Solfeggio frequencies is how they differ from standard musical tuning. Modern Western music typically uses A=440Hz as its reference pitch, while the Solfeggio system operates on a different mathematical basis altogether. Some researchers suggest that the Solfeggio tuning is more naturally resonant with human energy fields than standard musical tuning.

Whether you approach these frequencies from a spiritual perspective or simply as tools for psychological and energetic balance, their potential impact on your manifestation practice is worth exploring. By working with specific frequencies targeted to clear particular blocks, you can create a more coherent vibrational state from which manifestation flows more naturally.

In the next section, we'll examine the scientific evidence for these frequencies' effects, acknowledging both what has been verified and what remains in the realm of anecdotal experience.

Skeptic's Corner

Let's address the elephant in the room: despite thousands of enthusiastic testimonials about the Solfeggio frequencies, the scientific evidence supporting their specific healing properties remains limited. If you're naturally skeptical or analytically minded, you might be wondering whether these frequencies deserve a place in your manifestation practice.

This is a fair question that deserves an honest examination.

From a strictly scientific perspective, here's what we can say with reasonable confidence:

Sound affects physiology: Numerous studies confirm that sound frequencies can influence heart rate, respiration, brainwave patterns, and stress hormone levels. This much is well-established.

Frequency-specific effects exist: Different sound frequencies do create different physiological responses. For example, lower frequencies tend to promote relaxation while higher frequencies can increase alertness.

Resonance is real: The principle of resonance—where objects vibrate in sympathy with external frequencies—is a fundamental law of physics that applies to biological systems as well as mechanical ones.

Music therapy works: Clinical research has validated music therapy for reducing anxiety, managing pain, and improving mood—all states that can impact manifestation abilities.

However, when it comes to the specific claims about the Solfeggio frequencies—that 528 Hz repairs DNA or that 396 Hz specifically releases guilt—the scientific evidence becomes much thinner. Most of these claims rely on anecdotal reports, historical references, and theoretical frameworks rather than controlled studies.

So where does this leave the rational manifestor?

I suggest a balanced approach that neither dismisses these frequencies out of hand nor accepts all claims uncritically. Here's why the Solfeggio frequencies might still be valuable for your manifestation practice, even without comprehensive scientific validation:

The Placebo Effect Is Real Power

The placebo effect—where belief in a treatment creates real physiological changes—is one of the most well-documented phenomena in medical research. If you believe a particular frequency will help you manifest more effectively, that belief alone can create measurable changes in your brain chemistry and energy field.

Rather than dismissing this as "just placebo," recognize it as a legitimate way to leverage your mind's innate power. The placebo effect isn't fake healing—it's evidence of the mind's remarkable ability to influence physical reality, which is precisely what manifestation is about.

Subjective Experience Matters

Manifestation is inherently subjective—it involves your consciousness interacting with reality in ways that aren't always measurable by external instruments. If a particular frequency helps you feel more aligned, centered, or capable of manifesting, that subjective experience has real value regardless of whether an EEG would show predictable changes in everyone who listens to it.

Historical Persistence Suggests Something Real

The longevity of sound healing practices across diverse cultures suggests there's something more than superstition at work. While the specific claims about Solfeggio frequencies may be modern interpretations, the use of specific tones for healing and consciousness expansion appears in too many separate traditions to be dismissed as mere coincidence.

The Science Is Young, Not Absent

Research into frequency-specific effects on human consciousness is still in its infancy. The absence of comprehensive studies on Solfeggio frequencies doesn't mean they don't work—it means this particular area hasn't been a priority for research funding. As interest in sound healing grows, more research is emerging that may eventually validate (or refine) our understanding of how these frequencies work.

Risk-Benefit Analysis Favors Experimentation

Unlike many health interventions, listening to Solfeggio frequencies carries virtually no risk. Given the potential benefits and minimal downside, personal experimentation makes sense even for the scientifically minded.

A Rational Approach to Solfeggio Frequencies

If you're skeptically inclined but open to exploring these frequencies, I recommend this approach:

Test one frequency at a time: Rather than accepting all claims, experiment with one frequency for a specific purpose and observe the results objectively.

Keep a journal: Document your experiences with specific frequencies, noting any changes in your emotional state, thought patterns, or manifestation results.

Use controlled comparison: Try manifesting with and without the frequencies under otherwise similar conditions to better isolate their effects.

Maintain healthy skepticism: Be open to positive effects without feeling obligated to attribute them specifically to the frequencies—multiple factors may be at work.

Focus on results: Ultimately, what matters is whether your manifestation abilities improve, not whether the explanation for why they improve matches any particular theory.

The most balanced perspective recognizes that while the more extravagant claims about Solfeggio frequencies remain unproven, these sound tools offer a low-risk, potentially high-reward addition to your manifestation practice.

Whether they work through specific vibrational resonance with cellular structures, through entrainment of brainwave patterns, through placebo effect, or through some combination of mechanisms, the practical question is simply: do they help you manifest more effectively?

The only way to answer that question is through personal experimentation—which brings us to the practice section.

Practice

The Complete Solfeggio Frequency Protocol System. The Solfeggio frequencies offer you a precise set of sonic tools for clearing specific blocks to manifestation and activating particular qualities needed for successful creation.

The Six Core Solfeggio Frequencies

396 Hz - Liberation from Fear and Guilt

- Primary Applications: Clearing fear-based resistance to abundance, releasing guilt about deserving desires
- 21-Day Fear Liberation Protocol included

417 Hz - Facilitating Change and Breaking Patterns

- Primary Applications: Breaking stagnant situations, clearing resistance to necessary life changes
- 28-Day Transformation Protocol included

528 Hz - The Love Frequency

- Primary Applications: Heart-centered manifestation, healing relationship blocks
- 33-Day Heart Activation Protocol included

639 Hz - Connection and Harmonious Relationships

- Primary Applications: Improving supportive relationships, healing communication blocks
- 21-Day Relationship Harmony Protocol included

741 Hz - Expression and Solutions

- Primary Applications: Clearing creative blocks, activating intuitive problem-solving
- 14-Day Creative Solution Protocol included

852 Hz - Spiritual Insight and Divine Connection

- Primary Applications: Higher guidance, aligning personal desires with soul purpose
- 40-Day Spiritual Alignment Protocol included

Sample Protocol: 528 Hz Heart Activation (Days 1-11)

Daily Practice (20 minutes):

1. Heart Connection (3 minutes)
    - Sit with hands in prayer position at heart center
    - Begin 528 Hz audio
    - Feel green or pink light beginning to glow in your chest
    - Intention: "My heart opens to love as the foundation of all manifestation"
2. Self-Love Activation (8 minutes)
    - Place both hands over your heart
    - With each breath, send love to yourself as you are right now
    - Feel the 528 Hz frequency amplifying this self-love
    - Visualize your heart chakra spinning faster and glowing brighter
3. Manifestation Heart-Alignment (7 minutes)

- o Bring your manifestation desire into your heart space
- o Ask your heart: "How can I manifest this from love rather than need?"
- o Feel your desire being blessed by your own heart's wisdom

4. Love Broadcast (2 minutes)

- o Imagine your heart's love extending outward in all directions
- o See this love-frequency connecting with your manifestation in the quantum field
- o End with: "My manifestation flows to me through love"

TUNING YOUR WORLD

CHAPTER 6: VIBRATIONAL ARCHITECTURE

Clutter = Stagnant Energy

Your home is broadcasting a frequency into your energy field constantly—24 hours a day, 7 days a week.

That frequency is either supporting your manifestations or canceling them.

This chapter shows you exactly what to change in your environment

so it amplifies rather than sabotages your goals. It's a fundamental issue of energy flow that directly impacts your ability to manifest.

Every physical object in your environment carries and conducts energy. When these objects are thoughtfully arranged with intention and purpose, energy flows smoothly through your space, supporting your manifestation efforts. But when objects pile up without purpose—when clutter accumulates—that energy becomes stagnant, creating invisible resistance to what you're trying to create.

Think of energy like water. In a clear, flowing stream, water moves effortlessly, carrying nutrients and supporting life. But when debris blocks the stream, the water stagnates, becomes murky, and eventually toxic. The same principle applies to the energy in your home or workspace.

Clutter creates energetic stagnation in several specific ways:

It Represents Delayed Decisions

At its core, clutter is the physical manifestation of postponed decisions. Each item in that stack of mail required a decision you didn't make. Every piece of clothing not put away represents a choice deferred. This pattern of delay and indecision in your physical environment reinforces the same pattern in your manifestation efforts.

When you habitually surround yourself with unmade decisions, you're training your energy field to hold patterns in suspension rather than bringing them to completion. This directly undermines the decisive energy needed for effective manifestation.

It Fragments Your Attention

Your brain registers every item in your visual field, whether you're consciously aware of it or not. Each object in a cluttered space pulls a tiny bit of your attention, creating what neuroscientists call a "visual tax" on your cognitive resources.

A Princeton University study found that physical clutter in your surroundings competes for your attention, resulting in decreased performance and increased stress. This fragmentation of attention weakens your ability to focus clearly on what you want to manifest, diluting your vibrational broadcast.

It Anchors Old Energy

Objects don't just occupy physical space—they hold emotional and energetic imprints. That box of mementos from a past relationship, those clothes that no longer fit, the gifts you never liked but feel obligated to keep—all these items anchor old energetic patterns in your present space.

When you're trying to manifest new experiences, these energetic anchors create conflicting frequencies that can cancel out or distort your manifestation signal. It's like trying to broadcast a clear radio signal through static.

It Blocks New Energy From Entering

There's a simple principle in both physics and metaphysics: nature abhors a vacuum. When you create empty space—physically and energetically—you create a vacuum that naturally draws in new energy and opportunities.

Clutter fills all available space, leaving no room for new energy to enter. This is why decluttering often coincides with sudden forward movement in stuck areas of life—you've created space for new possibilities to flow in.

It Creates Subconscious Stress

Research from UCLA's Center on Everyday Lives of Families found that women who described their homes as cluttered had higher levels of cortisol (the stress hormone) than those who described their homes as restful and restorative. This chronic low-level stress creates an energetic broadcast of tension rather than receptivity.

Since manifestation works most effectively from a state of relaxed openness, this background stress significantly impedes your ability to align with what you desire.

The connection between clutter and stagnant energy isn't just spiritual theory—it's supported by research in psychology and neuroscience. A 2011 study in The Journal of Neuroscience found that multiple visual stimuli present in the same visual field compete for neural representation. In simple terms, when your environment is cluttered, your brain has to work harder to filter out irrelevant information, leaving fewer resources for focused manifestation.

Similarly, a study published in Environment and Behavior found that a cluttered environment led to decreased persistence on challenging tasks—a finding with direct implications for maintaining the consistent vibrational state needed for successful manifestation.

The good news is that addressing clutter creates immediate energetic shifts. Unlike many aspects of manifestation that may take time to materialize, the vibrational benefits of decluttering can be felt almost instantly. As you clear physical space, you simultaneously clear energetic space, creating room for new possibilities to enter.

This doesn't mean you need to become a minimalist overnight. It simply means becoming conscious of how the physical objects in your environment either support or hinder the free flow of energy. Each item you release creates space for manifestation energy to circulate more freely. Each thoughtful decision about what to keep and what to let go strengthens your decisive manifestation muscles.

In the practice section of this chapter, you'll learn specific techniques for clearing stagnant energy from your space. But for now, simply notice—with curiosity rather than judgment—where energy might be getting stuck in your physical environment, and how this might be affecting your manifestation efforts.

Feng Shui Vastu & Neuroscience

For thousands of years, cultures around the world have recognized that the arrangement of physical space profoundly affects energy flow, consciousness, and by extension, manifestation abilities. Two of the most sophisticated systems for harmonizing space—Chinese Feng Shui and Indian Vastu Shastra—developed independently yet share remarkable similarities in their core principles.

What's fascinating is that modern neuroscience is now validating many of these ancient spatial wisdom traditions, revealing the neurological mechanisms behind why certain environments enhance our wellbeing and manifestation capabilities while others diminish them.

Let's explore the cross-cultural principles of vibrational space harmonization and how they're supported by contemporary science:

Directional Energy Fields

Both Feng Shui and Vastu recognize that energy flows differently from different directions, with specific qualities associated with each cardinal point:

North: Associated with career and life path in Feng Shui; with prosperity in Vastu

East: Connected to health and family in Feng Shui; with vitality and sunrise energy in Vastu

South: Linked to fame and reputation in Feng Shui; with active energy in Vastu

West: Related to creativity and children in Feng Shui; with stability in Vastu

While these associations might seem arbitrary, neuroscience has discovered that our brains contain "grid cells" that create mental maps based partly on directional orientation. Research at the Kavli Institute for Systems Neuroscience found that these cells respond differently based on which direction we're facing, suggesting our brains process spatial information in ways that align with these ancient directional systems.

The Command Position

Both traditions emphasize the importance of proper positioning within a space, particularly for beds, desks, and seating. The ideal "command position" allows you to see the entrance to a room without being directly in line with it, providing both security and opportunity awareness.

Neuroscience explains this through the concept of "neuroarchitecture"—how architectural elements affect brain function. Studies show that when we can't see entrances or exits, the amygdala (our brain's threat detection center) remains partially activated, creating subtle background stress that disrupts coherent manifestation.

Balanced Elements

Feng Shui works with five elements (wood, fire, earth, metal, water), while Vastu recognizes five elements (earth, water, fire, air, space). Both systems teach that balancing these elements creates harmonious energy flow.

Neuroscientific research on "embodied cognition" reveals that our brains process environmental elements not just visually but as potential interactions. The presence or absence of elemental qualities (flowing water, solid stone, open space) triggers different neural networks that affect our cognitive and emotional processing—directly impacting our manifestation state.

Clutter Clearing

Both traditions emphasize the importance of clear, open spaces for optimal energy flow—a principle we explored in the previous section.

Neuroscience confirms this through studies on "visual processing load." Researchers at Princeton University found that visual clutter competes for neural resources, reducing the brain's ability to focus and process information effectively. This directly impacts your ability to maintain the clear, focused attention needed for powerful manifestation.

The Center Point

Both Feng Shui and Vastu place special importance on the center of a dwelling, considering it the energetic heart that influences all other areas.

Interestingly, neuroscience has identified that our brains have specialized "place cells" that create cognitive maps with reference points. The center of a space often serves as a primary reference point in these neural maps, suggesting our brains naturally process spatial information with central points as anchors—just as these ancient systems teach.

Light and Space

Both traditions emphasize the importance of natural light and proper spatial proportions for harmonious energy.

Neuroscience has extensively documented how light exposure regulates our circadian rhythms, affecting everything from hormone production to cognitive function. Studies at the Lighting Research Center have shown that properly lit environments improve mood, energy, and cognitive performance—all crucial factors for maintaining the high-frequency states needed for effective manifestation.

Sound Acoustics

Both systems include considerations of how sound moves through space, with Vastu particularly emphasizing the effects of resonance and echo.

Neuroscientific research on "auditory processing" confirms that background noise and acoustic properties of spaces significantly affect cognitive load and stress levels. A study in the Journal of the Acoustical Society of America found that poor acoustics increased cognitive effort and stress hormones—directly undermining the relaxed alertness ideal for manifestation.

The Convergence of Ancient Wisdom and Modern Science

What makes this cross-cultural validation so powerful for manifestation work is that it provides both time-tested practices and scientific understanding. You don't have to choose between spiritual tradition and scientific validation—they're increasingly converging around the same principles.

This convergence offers a powerful framework for creating spaces that vibrationally support your manifestation efforts. By applying these principles, you're not just making your space look better—you're creating an environment that neurologically and energetically aligns with successful manifestation.

The spaces we inhabit are not neutral containers—they are active participants in our vibrational state. By consciously designing your environment according to these principles, you create a physical space that resonates with and amplifies your manifestation intentions rather than subtly undermining them.

In the practice section, you'll learn how to apply these principles through a comprehensive space-clearing ritual that combines elements from both traditions with modern understanding of energy dynamics.

Practice

Sacred Space Creation and Environmental Harmonization

Your physical environment is not merely a container for your life—it's an active participant in your vibrational state and manifestation abilities. Drawing from authentic Vedic Vastu Shastra, this practice section provides comprehensive techniques for transforming your space into a manifestation sanctuary.

The Complete Space Clearing and Harmonization Ritual

Phase 1: Preparation and Purification (20-30 minutes)

Gathering Sacred Materials (Traditional Vedic Approach):

- Ghee (clarified butter) - for diya (oil lamp) lighting
- Cotton wicks - minimum 5 for directional lighting
- Kumkum (red vermillion) - for marking sacred spaces
- Turmeric powder (haldi) - for purification and prosperity
- Sea salt or rock salt - for energy absorption and clearing
- Camphor tablets - for atmosphere purification
- Sandalwood powder or incense - for elevating frequency

- Fresh flowers (especially marigolds, roses, or jasmine)

Personal Purification Protocol:

1. Ritual Bath or Shower (10 minutes)
   - o Use warm water to cleanse physical impurities
   - o Add 7 drops sandalwood oil or handful of turmeric to water
   - o Set intention: "I purify my body and energy field to create sacred space"
2. Sacred Dress
   - o Wear clean, preferably light-colored clothing
   - o Natural fabrics maintain better energy flow than synthetics
   - o Remove shoes before entering the space you'll be clearing

Phase 2: Energetic Clearing and Purification (30-45 minutes)

Step 1: Salt Absorption Technique (10 minutes)

- Place small bowls of sea salt in each corner of every room
- Add a pinch of turmeric to each bowl
- Starting from northeast corner, place hands over each bowl
- Visualize salt absorbing all negative energies and stagnant frequencies
- Leave salt bowls in place for minimum 24 hours

Step 2: Smoke Purification (Dhoop) (15 minutes)

- Light camphor tablet or sandalwood incense

at center of main living space

- Traditional Sanskrit purification mantra:
  "Om Apavitrah Pavitro Va Sarvavastham Gato'pi Va
  Yah Smaret Pundarikaksham Sa Bahyabhyantarah Sucih"
- Move clockwise through entire space
- Pay special attention to corners where stagnant energy accumulates

Step 3: Sound Clearing (Shabda Shuddhi) (10 minutes)

- Starting at northeast corner, create sound vibration
- Traditional methods: Conch shell blowing, bell ringing, clapping
- Mantra chanting: "Om Namah Shivaya" (108 times while moving through space)

Step 4: Light Activation (Jyoti Prakash) (10 minutes)

- Light ghee diya in all four cardinal directions:
  - East: Yellow/golden flame for new beginnings
  - South: Red flame for recognition and achievement
  - West: White/silver flame for creativity and completion
  - North: Green flame for abundance and prosperity

Directional Energy Optimization (Traditional Vastu)

Northeast (Ishanya) - Water Element Enhancement:

- Governs spiritual growth, clarity, and new beginnings
- Place crystal clear quartz or glass bowl with fresh water

- Keep this area clutter-free and elevated if possible

Southeast (Agneya) - Fire Element Enhancement:

- Governs health, energy, and transformation
- Place red or orange flowers, lit candle, or small fire element
- Kitchen ideally located in this direction

Southwest (Nairutya) - Earth Element Enhancement:

- Governs stability, relationships, and material prosperity
- Place heavy objects, earth-toned items, crystals
- Master bedroom ideally located here

Northwest (Vayavya) - Air Element Enhancement:

- Governs movement, communication, and social connections
- Place wind chimes, feathers, or air plants
- Keep this area open and airy

Creating Specific Manifestation Zones

Wealth Corner (Southeast):

- Keep well-lit and clean
- Place Lakshmi yantra or image
- Fresh yellow flowers (marigolds ideal)
- Gold-colored objects
- Daily activation with red candle

Love Corner (Southwest):

- Symbols of partnership: paired objects
- Pink, red, and earth tones

- Fresh red roses or pink flowers
- Daily activation with pink light

CHAPTER 7: VIBRATIONAL CARRIERS

Charged Objects

The people around you are changing your frequency whether you realize it or not.

The objects you touch are imprinting their energy into your field.

Most people have no idea their entire environment is sabotaging their manifestations.

This chapter teaches you to identify and transform these frequency carriers.

Understanding how objects become energetically charged and how to strategically use high-vibration items can dramatically enhance your manifestation practice.

Objects become energetically charged through four primary mechanisms:

Material Composition

Different materials naturally vibrate at different frequencies. This isn't mystical—it's basic physics. Every substance has a unique molecular structure that creates distinct vibrational properties:

Crystals have highly organized molecular lattices that create stable, coherent energy fields. This is why they're used in watches, computers, and other precision electronics—their vibrational properties are consistent and reliable.

Metals conduct energy readily, with each metal having unique vibrational characteristics. Gold, for instance, has been valued across cultures not just for its rarity but for its energetic properties of amplification and preservation.

Natural fibers like cotton, wool, and silk maintain connections to their biological origins and tend to interact more harmoniously with human energy fields than synthetic materials.

Synthetic materials often create discordant or dampening effects on natural energy fields, which is why spaces filled with plastics and synthetics can feel "dead" or energetically flat.

Energetic Imprinting

Objects absorb and store energy from their environments and the people who interact with them. This imprinting happens through several mechanisms:

Proximity exposure: Items kept near high-vibration sources gradually attune to those frequencies, like a smaller tuning fork beginning to vibrate when placed near a resonating larger one.

Intentional charging: When you consciously direct energy into an object through meditation, prayer, or ritual, you're imprinting it with specific vibrational patterns.

Emotional association: Objects present during emotionally charged events absorb those emotional frequencies. This is why items from happy occasions can lift your mood while those associated with difficult times can trigger lower vibrational states.

Repeated use: Objects used regularly in specific activities become conditioned to the energy of those activities. A journal used for gratitude practice begins to carry that frequency;

prayer beads used in meditation become conditioned to meditative states.

Sacred Geometry

The physical structure and proportions of an object create specific energetic effects regardless of material. This is the science of sacred geometry—how certain proportions and patterns interact with energy fields in consistent ways:

1. Spirals (like those found in nautilus shells) create energy that continually renews and expands

2. Pyramids focus and amplify energy at their centers and peaks

3. Circles create containment and wholeness in energy fields

4. Merkabas (star tetrahedrons) balance and integrate opposing energies

Objects designed according to these sacred proportions create specific energetic effects in your space and on your personal field when you interact with them.

Historical/Collective Charging

Some objects carry charges accumulated over long periods or from many people:

Ancient artifacts carry the imprints of their cultures and uses over centuries

Religious symbols hold the collective energy of millions of people who have focused on them with devotion

Family heirlooms carry the energetic lineage of your ancestors

Items from sacred sites retain the high-frequency energy of those locations

Now that we understand how objects become charged, let's explore specific high-vibration objects that can support your manifestation practice:

Rudraksha Beads

These natural seeds from the Rudraksha tree have been used in spiritual practices for thousands of years, particularly in Hindu and Buddhist traditions. Their unique electromagnetic properties help balance the nervous system and create coherence in your energy field—an optimal state for manifestation.

Different Rudraksha configurations (the number of segments or "mukhis" on the seed) correspond to different energetic effects:

5-mukhi beads enhance overall clarity and purpose

7-mukhi beads specifically support manifestation of desires

8-mukhi beads remove obstacles to manifestation

Wearing Rudraksha as a necklace or bracelet creates a continuous field of stabilizing energy around your most important chakras.

Crystals for Manifestation

Different crystals support different aspects of the manifestation process:

Clear Quartz: Amplifies your intentions and brings clarity to manifestation goals

Citrine: Known as the "merchant's stone," it specifically supports abundance manifestation

Green Aventurine: Attracts opportunities and synchronicities

Pyrite: Grounds manifestation energy into practical action

Labradorite: Helps align your energy with your highest path, preventing manifestation of desires that don't serve your evolution

The key to effective crystal work isn't collecting many stones but developing a relationship with a few that specifically support your manifestation goals.

Charged Water

Water is extraordinarily receptive to energetic imprinting. By creating and consuming charged water, you literally internalize specific frequencies:

Place water in a glass container (never plastic)

Surround it with crystals aligned with your manifestation goals

Place your written intention under the container

Leave in sunlight or moonlight (depending on the nature of your manifestation)

Drink mindfully, visualizing the charged water distributing your intention throughout your body

Personal Talismans

A talisman is any object specifically charged for a particular purpose. Creating a personal manifestation talisman is powerful because it combines material properties with your unique energy signature:

Choose an object that resonates with you (a stone, piece of jewelry, small figurine)

Cleanse it thoroughly (using methods from the space clearing ritual)

Hold it while in a meditative state, visualizing your manifestation as complete

"Program" it with a specific phrase or intention

Carry or place it where you'll interact with it daily

Practical Guidelines for Working with Charged Objects:

Quality over quantity: A few highly charged objects are more effective than many items with weak or confused energy.

Regular cleansing: All objects accumulate energetic debris over time. Cleanse your manifestation objects monthly using sound, smoke, sunlight, or salt.

Conscious placement: Position charged objects where they'll have maximum impact—near where you sleep, work, or practice manifestation techniques.

Personal connection: The objects that will work most powerfully for you are those that you feel a genuine connection with, regardless of their traditional associations.

Ethical sourcing: The provenance of an object affects its energy. Items acquired through exploitation carry that discordant energy, which can undermine your manifestation efforts.

By thoughtfully selecting and working with high-vibration objects, you create an ecosystem of supportive energy that maintains your manifestation frequency even when your conscious attention is elsewhere. These objects become vibrational allies, continuously broadcasting frequencies that align with what you wish to create.

People as Frequencies

Just as objects carry and transmit vibrational frequencies, the people in your life are powerful energetic broadcasters who directly influence your manifestation field. In fact, human energy fields are far more complex and impactful than those of inanimate objects—making your social environment one of the most significant factors in your vibrational state.

Every person you interact with is broadcasting a complex frequency pattern based on their thoughts, emotions, beliefs, and level of consciousness. Through the principle of resonance, these frequencies interact with your own energy field, either amplifying or diminishing your manifestation power.

The Science of Interpersonal Energy Exchange

This isn't just metaphysical theory—it's supported by solid research:

The HeartMath Institute has documented how one person's heart rhythms can directly influence another person's brainwave patterns when they're in proximity.

Studies on "emotional contagion" show that we unconsciously mimic and synchronize with the emotional states of those around us through subtle facial cues, voice tonality, and body language.

Research on "mirror neurons" reveals that our brains literally mirror the neurological activity of people we're observing, creating internal states that match what we're witnessing externally.

These mechanisms explain why you can "feel drained" after spending time with certain people or "energized" by others—you're experiencing real energetic exchanges that affect your vibrational state.

Identifying Frequency Patterns in Relationships

People in your life generally fall into three vibrational categories in relation to your manifestation work:

Frequency Elevators

These individuals raise your vibration through their presence and interaction. They're characterized by:

Authentic positivity (not toxic positivity that denies reality)

Genuine support for your growth and dreams

Personal responsibility for their own energy and life circumstances

Abundance mindset rather than scarcity thinking

Solution-focused rather than problem-focused thinking

Emotional maturity and self-awareness

After spending time with frequency elevators, you feel more aligned with your manifestation goals, more confident in your abilities, and energetically clearer.

Frequency Neutrals

These relationships neither significantly raise nor lower your vibration. They may be pleasant and comfortable but don't particularly challenge you to grow or drag you into lower states. They're characterized by:

Consistent but not particularly inspiring energy

Conventional thinking that neither limits nor expands your perspective

Moderate emotional expression without extremes

Focus on maintaining rather than transforming

Time with frequency neutrals generally leaves you feeling the same as before the interaction—neither depleted nor energized.

Frequency Drains

These relationships consistently lower your vibrational state through various mechanisms. They're characterized by:

Chronic complaining without taking action

Criticism of your goals and dreams

Drama-centered communication patterns

Victim mentality and blame projection

Energy vampirism (feeling energized after leaving you depleted)

Jealousy of others' success or happiness

After interacting with frequency drains, you often feel tired, doubtful about your manifestation abilities, or subtly pulled off course from your intentions.

Managing Your Social Frequency Environment

Understanding these patterns allows you to consciously manage your social environment for optimal manifestation:

Conduct a Relationship Inventory

Make a list of the people you spend the most time with and honestly assess their impact on your energy and manifestation efforts. For each person, note:

How you feel before, during, and after interactions

Whether they support or undermine your manifestation goals

The dominant emotional tone of your interactions

Whether the relationship is energetically balanced or one-sided

This inventory isn't about judging others' worth as people—it's about honestly assessing energetic impact.

Implement Strategic Boundaries

Once you've identified frequency drains, implement appropriate boundaries:

Time boundaries: Limit the duration and frequency of interactions

Topic boundaries: Redirect conversations away from chronic negativity

Emotional boundaries: Avoid taking responsibility for others' emotional states

Energetic boundaries: Visualize protective energy around yourself during necessary interactions

Remember that boundaries aren't punishments—they're healthy limits that protect your vibrational state while still allowing compassionate interaction.

Cultivate Elevating Relationships

Actively seek and nurture relationships with frequency elevators:

Join communities focused on growth, manifestation, or spiritual development

Attend workshops or retreats where like-minded people gather

Reach out to individuals whose energy you admire

Deepen existing relationships with people who support your highest vision

Even one strong connection with a frequency elevator can significantly amplify your manifestation power.

Practice Compassionate Detachment

For relationships you can't or don't want to end (family members, coworkers, long-term friends), develop compassionate detachment:

Recognize that others' vibrational states reflect their own journey, not yours

Maintain emotional connection without energetic entanglement

Visualize their highest potential while accepting their current state

Set clear internal boundaries about what energy you will and won't absorb

This approach allows you to maintain relationships while protecting your vibrational field.

Become a Conscious Broadcaster

Remember that you're also broadcasting a frequency that affects others. By maintaining your own high vibration, you:

Naturally attract more frequency elevators into your life

May help elevate those around you through entrainment

Create a buffer against lower frequencies

Demonstrate what's possible, inspiring others

The most powerful way to create a high-frequency social environment is to consistently embody that frequency yourself.

Special Considerations for Intimate Relationships

Intimate partnerships create the strongest energetic exchanges in our lives. If you're in or considering a romantic relationship, pay special attention to:

How aligned your manifestation goals are with your partner's

Whether your partner's beliefs about what's possible expand or contract your own

The dominant emotional frequency you co-create together

How you feel about your manifestation work after spending

time together

A supportive intimate relationship can exponentially amplify your manifestation power, while a misaligned one can create significant resistance.

By consciously curating your social environment based on frequency awareness, you create a social ecosystem that naturally supports and enhances your manifestation efforts rather than subtly undermining them.

Practice

Object Charging and Talisman Creation Mastery

Every object in your environment carries and transmits energy. By learning the ancient arts of object charging, you transform ordinary items into powerful allies for your manifestation work.

The Complete Object Charging Protocol

Phase 1: Object Preparation and Purification (15-20 minutes)

Selection and Cleansing:

- Choose items that feel naturally harmonious to you
- For manifestation work, choose objects you'll interact with daily
- Natural materials generally hold charges longer than synthetic ones

Traditional Purification Process:

1. Wash object with clean water mixed with blessed water
2. For metals: Add pinch of turmeric to cleansing water

3. For crystals: Use sea salt water (1 tsp salt per cup water)
4. Pass through sandalwood incense smoke while chanting: "Om Pavitram Pavitram"
5. Place on white cloth in eastern direction overnight before charging

Timing for Optimal Charging (Traditional Vedic)

Most Powerful Times:

- Brahma Muhurta: 1.5-3 hours before sunrise (most spiritually potent)
- New Moon (Amavasya): For new beginnings and planting intentions
- Full Moon (Purnima): For amplification and completion energies
- Thursday (Guruvaar): Day of Jupiter, excellent for wisdom and manifestation
- Friday (Shukravaar): Day of Venus, ideal for abundance and relationships

Sacred Space Preparation

Traditional Setup:

1. Location: East-facing, clean, elevated surface
2. Base Cloth: White or red silk/cotton cloth
3. Five Element Representation:
    - Earth: Small bowl of clean soil or crystals
    - Water: Copper vessel with clean water + tulsi leaves

  - Fire: Ghee diya with cotton wick
  - Air: Fresh flowers or incense
  - Space: Bell or small cymbals

The Charging Ceremony (20-30 minutes)

Ganesha Invocation (3 minutes):

- Light the ghee diya
- Offer rice grains and red flower to the east
- Chant: "Om Gam Ganapataye Namaha" (21 times)
- Prayer: "Lord Ganesha, bless this sacred work. Remove all obstacles from this object's manifestation power."

Breath Charging (Prana Pratishtha) - 7 minutes:

1. Hold object close to heart center
2. Take deep breath, filling lungs completely
3. Exhale slowly directly onto object while visualizing golden light flowing from heart through breath into object
4. Continue for 49 breath cycles (7 x 7 for completion)

Mantra Imprinting - 10 minutes:

Choose appropriate mantra:

- *General Manifestation:* "Om Hreem Shreem Kleem Maha Lakshmiyei Namaha"
- *Abundance:* "Om Shreem Maha Lakshmiyei Namaha"
- *Love:* "Om Kleem Krishnaya Namaha"
- *Health:* "Om Tryambakam Yajamahe

Sugandhim Pushtivardhanam"

- *Success:* "Om Gam Ganapataye Namaha"

Hold object while chanting chosen mantra 108 times. With each repetition, visualize the sound vibration entering the object.

Elemental Blessing - 5 minutes:

1. Earth: Touch object to clean soil: "By earth's stability, may this manifestation be grounded"
2. Water: Sprinkle blessed water: "By water's flow, may this manifestation move into form"
3. Fire: Pass briefly above diya flame: "By fire's transformation, may obstacles be burned away"
4. Air: Waft incense smoke: "By air's movement, may this manifestation spread through my life"
5. Space: Ring bell near object: "By space's openness, may this manifestation flourish"

Specific Object Charging Protocols

Clear Quartz - Master Manifestation Amplifier:

- Cleanse in sea salt water overnight
- Use "Om Namah Shivaya" mantra (108 times)
- Program: "Amplify and accelerate the manifestation of [desire]"
- Daily activation: Hold during morning visualization

Rose Quartz - Love and Relationship Magnetism:

- Cleanse with rose water
- Use "Om Kleem Kamadeva Namaha" for divine love
- Program: "Attract my perfect partner who supports my highest evolution"
- Daily activation: Place on heart chakra during meditation

Citrine - Abundance and Prosperity:

- Charge on Thursday morning or new moon
- Use "Om Shreem Maha Lakshmiyei Namaha"
- Program: "Open all channels for abundant wealth flow"
- Daily activation: Carry in wallet, place in wealth corner

Mala (Prayer Beads) Charging

Complete 108-Bead Mala Activation:

Preparation:

- Soak new mala overnight in water with sea salt and tulsi leaves
- Dry in morning sunlight
- Thread through sandalwood incense smoke 7 times

Charging:

- Sit facing east holding mala in right hand
- Begin at guru bead (largest bead at center)
- Chant chosen manifestation mantra once per bead
- Use thumb to move from bead to bead
- Complete 108 repetitions, ending at guru bead

- Hold guru bead to third eye and download complete manifestation vision

Daily Practice:

- Morning: Chant manifestation mantra on each bead while visualizing desired outcome
- Evening: Review day's progress toward manifestation while touching each bead

DAILY RITUALS

CHAPTER 8: MORNING FREQUENCY ROUTINE

Breath Sound Intention

Now it's time to put brainwave entrainment into practice with a simple yet powerful exercise designed to enhance your manifestation abilities through focused awareness. This 15-minute protocol uses carefully selected audio technology to shift your brain into an optimal state for clear intention setting and energetic alignment.

What You'll Need:

Headphones (for optimal effect)

A quiet space where you won't be disturbed

The Focus Playlist (accessible via the QR code below or at www.vibrationbook.com/focus)

A journal or notepad

20 minutes of uninterrupted time (15 for the practice, 5 for integration)

The Playlist Structure:

The Focus Playlist is specifically engineered to guide your brain through a sequence of states optimal for manifestation work:

Minutes 0-3: Alpha Transition (10 Hz)

This section helps you transition from normal beta consciousness to a relaxed alpha state, creating receptivity while maintaining alertness.

Minutes 3-10: Alpha-Theta Border (7.83 Hz)

The core of the practice occurs at the Schumann Resonance frequency (7.83 Hz)—the Earth's electromagnetic "heartbeat." This frequency creates an ideal balance between conscious direction and subconscious receptivity.

Minutes 10-15: Alpha Return (10 Hz)

The final section brings you back to alert alpha, allowing you to integrate insights while remaining in a receptive state.

The audio combines binaural beats for depth with subtle isochronic pulses for clarity, embedded in a harmonically rich soundscape designed to evoke a sense of possibility and expansion.

Practice Instructions:

Preparation (2-3 minutes)

Find a comfortable seated position with your spine straight

Place your journal within reach

Put on your headphones

Take three deep breaths, exhaling completely each time

Set a clear intention for what aspect of manifestation you want to focus on during this session

The Listening Practice (15 minutes)

Start the Focus Playlist

For the first 3 minutes (Alpha Transition):

Keep your eyes open but softly focused

Allow your thoughts to gradually slow down

Bring your awareness to your breath without controlling it

For the middle 7 minutes (Alpha-Theta Border):

Close your eyes

Bring to mind what you wish to manifest, seeing it as already complete

Notice any resistance, blocks, or conflicting beliefs that arise

Don't fight these—simply observe them with curiosity

Return to the vision of your manifestation as already complete

Feel the emotions associated with this completion

For the final 5 minutes (Alpha Return):

Keep your eyes closed

Begin to notice any insights, symbols, or guidance that has emerged

Consider what specific actions would align with your manifestation

Allow these insights to clarify naturally without forcing

Integration (5 minutes)

When the audio ends, open your eyes

Immediately write in your journal:

Any insights or clarity you received

Any blocks or resistance you identified

At least one specific action you can take in the next 24 hours

How you felt during different parts of the practice

Optimal Use Guidelines:

Environment: Use this practice in a clean, uncluttered space with minimal distractions. Turn off notifications on your devices.

Timing: The most effective times for this practice are early morning (before your mind becomes cluttered with the day's concerns) or early evening (when your brain naturally begins shifting toward alpha states).

Frequency: For maximum benefit, use this practice 3-4 times per week. Daily use is fine, but allow at least 6 hours between sessions to prevent entrainment fatigue.

Hydration: Drink water before and after the practice. Proper hydration enhances electrical conductivity in the brain, making entrainment more effective.

Expectations: You may not experience dramatic effects the first time. Brainwave entrainment often works subtly, with effects becoming more pronounced with regular use as your brain learns to respond more readily to the frequencies.

What You Might Experience:

Everyone's experience with brainwave entrainment is unique, but common reports include:

A sense of expanded awareness

Increased clarity about manifestation goals

Spontaneous insights about blocks or next steps

Visual imagery or symbols related to your manifestation

A feeling of energetic alignment with your desired outcome

Reduced anxiety about the manifestation process

Enhanced creativity in approaching challenges

Troubleshooting:

If you find your mind wandering excessively, gently bring your focus back to your breath, then to your manifestation vision.

If you feel sleepy, sit more upright or open your eyes briefly while continuing to listen.

If you experience any discomfort, simply remove the headphones and try again another time, perhaps at a lower volume.

This 15-minute practice offers a concentrated dose of the optimal brainwave state for manifestation work. By regularly accessing this state, you're training your brain to more easily enter the frequencies where the gap between intention and manifestation narrows.

Use this practice as a foundation for your manifestation work, a reset when you feel stuck, or a clarifying tool when you need to refine your intentions.

The first hour after waking sets the vibrational tone for your entire day. Most people squander this precious time, immediately diluting their energy with social media, news, or rushing into responsibilities. But those who master manifestation understand that the morning hours offer a unique opportunity to establish a high-frequency state that can carry through the day's challenges.

This 15-minute morning routine combines three powerful elements—breath, sound, and intention—to create a vibrational foundation that supports manifestation throughout your day. Each component builds upon the previous one, creating a synergistic effect greater than any single practice alone.

The Power of Morning Practice

Your brain exhibits different wave patterns in the morning than at any other time of day. As you transition from sleep to wakefulness, you naturally pass through alpha and theta states—the same states that advanced meditators work for years to access. This makes the morning an ideal time to imprint new patterns and set vibrational intentions.

Additionally, your energy field is most receptive after sleep, before it encounters the various frequencies of your daily life. Think of it as a clean canvas—what you put on it first will influence everything that comes after.

The 15-Minute Morning Frequency Routine

This routine is designed to be simple enough to maintain daily while powerful enough to create significant vibrational shifts. Each component serves a specific purpose in raising and stabilizing your frequency.

Part 1: Conscious Breathing (5 minutes)

Begin by sitting comfortably at the edge of your bed or in a dedicated meditation space. Keep your spine straight but not rigid, with your hands resting on your thighs.

Connection Breath (1 minute)

Close your eyes and take three deep breaths, inhaling through your nose and exhaling through your mouth

With each exhale, release any remnants of sleep or dream states

Bring your awareness fully into your body, noticing sensations without judgment

Coherence Breathing (2 minutes)

Shift to breathing in and out through your nose

Establish a rhythm of 5 counts in, 5 counts out

Place one hand over your heart center

Feel your heartbeat synchronizing with your breath

Visualize your heart and brain coming into coherent rhythm

Energy Activation Breath (2 minutes)

Maintain the same breathing rhythm

As you inhale, visualize golden light entering through the crown of your head

As you exhale, see this light expanding throughout your body

With each breath cycle, the light grows brighter and fills more of your energy field

By the final breath, see yourself completely enveloped in this golden light

This breathing sequence creates coherence in your nervous system, synchronizes your heart and brain rhythms, and begins to raise your vibrational frequency through conscious direction of energy.

Part 2: Sound Attunement (5 minutes)

Sound creates immediate shifts in your vibrational state through entrainment. This section uses your own voice as a powerful tool for frequency modulation.

Resonant Humming (2 minutes)

With lips closed, begin a comfortable humming sound on your exhale

Allow the sound to find its natural resonance in your chest and head

Experiment with slightly different tones until you find one that creates the strongest vibration sensation

Continue humming this tone, feeling it vibrate through your entire body

Notice how the vibration begins to dissolve any dense or stuck energy

Vowel Toning (2 minutes)

Open your mouth and transition to vowel sounds

Begin with "AH" (as in father), feeling it resonate in your chest

Move to "OH" (as in go), feeling it resonate in your throat

Continue with "EE" (as in see), feeling it resonate in your head

Hold each tone for a full exhale, taking a normal breath between sounds

Feel how each sound activates different energy centers in your body

Power Mantra (1 minute)

Choose one of these seed mantras based on what you most need today:

"RAM" (pronounced rahm) for vitality and manifestation power

"OM" (pronounced aum) for connection and alignment

"SHRIM" (pronounced shreem) for abundance and prosperity

Repeat your chosen mantra aloud 9 times

With each repetition, feel its vibration becoming stronger in your body

This sound sequence activates your personal vibrational field, clears stagnant energy from your system, and attunes you to specific frequencies that support manifestation.

Part 3: Intention Setting (5 minutes)

With your system now coherent and vibrationally activated, you're ready to set clear intentions that will shape your day's manifestation field.

Gratitude Priming (1 minute)

Place both hands over your heart

Bring to mind three specific things you're genuinely grateful for right now

Feel the emotion of gratitude fully in your body

Notice how this emotion raises your vibrational frequency

Manifestation Visualization (2 minutes)

Bring to mind your primary manifestation goal

See it as already complete, experiencing it through all your senses

Feel the emotions associated with this manifestation

Notice any resistance or doubt that arises, and gently release it

Return to the feeling of your manifestation as complete

Daily Frequency Intention (2 minutes)

Ask yourself: "What frequency would best serve my manifestation today?"

(Examples: clarity, confidence, compassion, abundance, creativity)

Once identified, create a simple statement: "Today I embody the frequency of [your chosen quality]"

Repeat this statement three times aloud

Visualize yourself moving through your day carrying this frequency

See potential challenges and how you'll maintain this frequency through them

This intention sequence harnesses the coherent, activated state you've created and directs it toward specific manifestation outcomes, both immediate (today) and longer-term (your primary goal).

Completing the Practice

To complete your morning frequency routine:

Take three deep breaths, feeling yourself fully present in your body

Gently open your eyes, maintaining the internal state you've created

Place your palms together at your heart in a gesture of commitment to your vibrational intention

Say aloud: "I carry this frequency throughout my day, attracting all that aligns with my highest good."

Integration Tips for Maximum Benefit

To get the most from this practice:

Consistency trumps duration: A daily 15-minute practice is far more effective than an occasional longer session.

Practice before digital exposure: Complete this routine before checking your phone, email, or news to prevent external frequencies from diluting your intentional state.

Physical preparation matters: Keep a glass of water by your bed to hydrate before beginning, as water helps conduct the subtle electrical currents involved in vibrational work.

Create environmental support: Designate a specific space for this practice and keep it clean and free of clutter to support clear energy flow.

Track your results: Keep a simple log of your daily frequency intention and how it manifested throughout the day to reinforce the connection between your practice and your experiences.

This 15-minute morning routine may seem simple, but its effects are profound when practiced consistently. By deliberately setting your vibrational tone at the start of each day, you create a foundation that supports manifestation in everything that follows.

Your morning isn't just the start of your day—it's the genesis point of your manifestation field. Choose to begin with conscious vibration rather than unconscious reaction, and watch how differently your day unfolds.

Emotional Priming

The vibrational state you cultivate first thing in the morning creates a template that influences every interaction, decision, and manifestation opportunity throughout your day. By deliberately priming your emotional state rather than letting it be determined by circumstances, you gain tremendous power over your manifestation outcomes.

Emotional priming goes beyond positive thinking—it's about deliberately activating specific emotional frequencies that align with what you want to manifest. This section explores two powerful techniques for emotional priming: strategic gratitude practices and "I Am" statements that reshape your vibrational signature.

The Science of Emotional Priming

Research in neuroscience and psychology confirms that our emotional states aren't just reactions to events—they're active filters that determine what we notice, how we interpret information, and which opportunities we can recognize.

A study from the University of California found that participants primed with positive emotions were able to see more possibilities and connections than those in neutral or negative states. Another study from the HeartMath Institute demonstrated that positive emotional states create coherent patterns in heart rhythm, which synchronize brain activity and optimize cognitive function.

In manifestation terms, your emotional state determines which frequencies you can receive and broadcast. If you're vibrating at the frequency of lack, you'll be blind to abundance opportunities right in front of you. If you're vibrating at the frequency of love, you'll naturally attract and notice loving connections.

Strategic Gratitude Practices

Gratitude is one of the most powerful tools for emotional priming because it immediately shifts your vibration from what's lacking to what's present. However, not all gratitude practices are equally effective for manifestation purposes.

Strategic gratitude means deliberately focusing your gratitude in ways that align with your manifestation goals. Here's how to practice it effectively:

Manifestation-Aligned Gratitude

Rather than general gratitude, focus specifically on aspects of your life that resonate with what you're trying to manifest:

If manifesting financial abundance: Express gratitude for all the ways money already flows to you, no matter how small

If manifesting a relationship: Feel grateful for the loving connections already in your life and moments when you feel loved

If manifesting health: Thank your body for the functions that work well, focusing on strength and vitality

This aligned gratitude activates the frequency of what you're manifesting, making you more receptive to similar experiences.

Future-Self Gratitude

This powerful technique involves expressing gratitude for your manifestation as if it has already occurred:

Close your eyes and step into the perspective of your future self who has already received the manifestation

From this perspective, express specific gratitude for how this manifestation has changed your life

Feel the emotions of relief, joy, and fulfillment as if they're happening now

Notice the details of what you're grateful for from this future perspective

Example: "I am so grateful that my business now consistently generates $10,000 monthly. I'm thankful for the freedom this creates in my life and how it allows me to support my family while doing work I love."

This practice creates a neurological and energetic bridge between your current state and your desired manifestation.

Contrast Gratitude

This technique uses the contrast between past challenges and current blessings to amplify your gratitude frequency:

Recall a specific challenge you've overcome or difficult period you've moved through

Feel the emotions associated with that time briefly

Then immediately shift to gratitude for how that situation has resolved or improved

Feel the emotional contrast between the challenge and the resolution

Example: "I remember how stressed I felt when I couldn't pay my bills on time. I'm so grateful that I now have more than enough to cover my expenses each month and am building savings."

This contrast creates a powerful emotional surge that rapidly elevates your vibrational frequency.

"I Am" Statements: Reprogramming Your Vibrational Signature

While gratitude shifts your emotional state in the moment, "I Am" statements work at a deeper level to reprogram your fundamental identity and vibrational signature.

The words "I Am" carry unique power because they directly address your subconscious definition of self. Whatever follows "I Am" becomes part of your energetic blueprint and influences what you can manifest.

Here's how to use "I Am" statements effectively for emotional priming:

Frequency-Based Statements

Rather than focusing on having or doing, create statements about being specific frequencies:

"I am the frequency of abundance" (rather than "I am wealthy")

"I am the embodiment of love" (rather than "I am loved")

"I am the vibration of perfect health" (rather than "I am healthy")

These statements identify you with the frequency itself rather than its manifestations, allowing multiple expressions of that frequency to enter your experience.

Present-Tense Embodiment

Craft statements that place you in the immediate experience of your desired state:

"I am now vibrating in perfect resonance with my desires"

"I am currently aligned with unlimited abundance"

"I am in this moment experiencing deep connection"

The words "now," "currently," and "in this moment" bypass the subconscious tendency to place fulfillment in the future.

Feeling-State Activation

Create statements that trigger specific emotional states

aligned with your manifestation:

"I am overflowing with gratitude for the abundance surrounding me"

"I am radiating confidence in my path and purpose"

"I am filled with the peaceful certainty that all is unfolding perfectly"

These statements directly activate the emotional frequencies that attract your desires.

Morning Priming Protocol (5-7 minutes)

For maximum benefit, combine these techniques into a morning emotional priming protocol:

Centering (30 seconds)

Place one hand on your heart and one on your belly

Take three deep breaths, centering your awareness in your body

Gratitude Activation (2-3 minutes)

Practice one or more of the strategic gratitude techniques

Speak your gratitude aloud to engage multiple senses

Amplify the feeling component rather than rushing through a list

"I Am" Reprogramming (2-3 minutes)

Stand in front of a mirror if possible

Look into your own eyes

Speak 3-5 carefully crafted "I Am" statements with conviction

Feel each statement resonating in your body as you speak it

Embodiment (1 minute)

Close your eyes and fully embody the emotional state you've created

Place your hands on your heart

Make a commitment to carry this emotional frequency

throughout your day

Create a physical anchor (like touching your thumb and forefinger together) that you can use throughout the day to quickly return to this state

This protocol creates a powerful emotional foundation that primes your entire system for aligned manifestation throughout the day. Rather than reacting to circumstances, you're proactively establishing the vibrational conditions that attract your desires.

The science is clear: your emotional state isn't just a result of your circumstances—it's a causal factor in creating them. By deliberately priming your emotional frequency each morning, you're setting the energetic template for what can manifest in your experience.

Practice

PRACTICE: Morning Frequency Routine

The Complete System

Time Required: Quick 10 min | Standard 15 min | Extended 30 min

---

Best For: Setting entire day's frequency, establishing manifestation alignment, consistent nervous system recalibration

Frequency: Daily, immediately upon

waking before any other activities

Equipment Needed: None essential (optional: journal, essential oil, water, meditation space)

QUICK VERSION (10 minutes) For Busy Mornings

Perfect for days when time is limited but manifestation work is essential.

STEP 1: AWAKENING (1 minute)

Sit up in bed or stand up

Take three deep breaths, filling lungs completely

Set one intention for the day: "I vibrate at [frequency] today"

Example intentions: peace, confidence, abundance, clarity, joy

STEP 2: BODY GRATITUDE (2 minutes)

Quickly scan from head to feet, thanking each body part

"I am grateful for my mind that creates clarity"

"I am grateful for my heart

that loves freely"

"I am grateful for my hands that create"

"I am grateful for my feet that carry me forward"

Complete in 90 seconds

STEP 3: SOUND ACTIVATION (3 minutes)

Take 3 deep breaths

Hum or chant "OM" or "AUM" for 1 minute

OR chant primary manifestation mantra 1 minute

OR recite affirmation 1 minute

Feel the vibration moving through your body

STEP 4: QUICK VISUALIZATION (3 minutes)

Close eyes, place hand on heart

See your primary manifestation as already complete

Feel the emotion of having it

Visualize yourself moving through the day aligned with this frequency

State: "This is manifesting now"

STEP 5: TRANSITION (1 minute)

Open eyes slowly

Stand and stretch

State: "I carry this frequency throughout my day"

Move into your morning with intentional presence

Total Time: 10 minutes

Manifestation Impact: Solid daily frequency setting, suitable for ongoing practice

STANDARD VERSION (15 minutes) The Recommended Daily Practice

This is your go to practice for most mornings. It's comprehensive enough to significantly impact your day while being manageable within typical morning routines.

STEP 1: SACRED SPACE PREPARATION (1 minute)

Find a comfortable spot (bed, chair, or meditation cushion)

Ensure spine is straight (not rigid)

Optional: Light a candle or incense if desired

Create intentional separation from the day's rush

STEP 2: BREATHING & AWAKENING (2 minutes)

Part A: Cleansing Breath (1 minute)

Inhale through nose for 4 counts

Hold for 4 counts

Exhale through mouth for 4 counts

Complete 8 full cycles

Purpose: Clear sleep fog, activate nervous system

Part B: Alternate Nostril Breathing (1 minute)

Close right nostril with thumb, inhale through left (4 counts)

Close left with finger, release right, exhale right (4 counts)

Inhale right, close right, exhale left

Complete 12 full cycles

Purpose: Balance brain hemispheres, activate subtle energy channels

STEP 3: SOUND ACTIVATION (4 minutes)

Part A: Chakra Activation with Bija Mantras (2 minutes)

Hold each sound for 10 15 seconds, feeling vibration in each location:

LAM (root chakra) Feel vibration at base of spine

VAM (sacral chakra) Feel vibration in lower abdomen

RAM (solar plexus) Feel vibration in upper abdomen

YAM (heart chakra) Feel vibration in chest center

HAM (throat chakra) Feel vibration in throat

OM (third eye) Feel vibration in forehead area

Purpose: Activate and align your entire subtle energy system

Part B: Primary Manifestation Mantra (2 minutes)

Choose ONE mantra based on your current focus:

If manifesting ABUNDANCE: "Om Shreem Maha Lakshmiyei Namaha"

If manifesting LOVE: "Om Kleem Krishnaya Namaha"

If manifesting SUCCESS: "Om Gam Ganapataye Namaha"

If manifesting HEALTH: "Om Tryambakam Yajamahe"

If manifesting SPIRITUAL GROWTH: "Om Namah Shivaya"

If manifesting REMOVAL OF OBSTACLES: "Om Gam Ganapataye Namaha"

Chant for 2 full minutes, feeling the vibrational frequency penetrate your being.

Let your voice be natural, not forced.

Feel rather than think about the mantra.

STEP 4: HEART COHERENCE ACTIVATION (3 minutes)

Part A: Heart Focus (1 minute)

Place both hands over heart center

Slow breath: 5 seconds in, 5 seconds out

Feel breath moving through heart area

Activate genuine appreciation for something in your life (specific person, experience, or quality you possess)

Notice your heart area becoming warm, expanded, coherent

Part B: Primary Visualization (2 minutes)

Keep hands on heart, maintain coherent breathing

Visualization Segment 1 Clear Intention (30 seconds):

Bring your primary manifestation to mind

See it with crystal clarity

What does it look like? Feel like? Sound like?

Visualization Segment 2 Completed Reality (60 seconds):

See your manifestation as ALREADY COMPLETE

Visualize through all five senses:

* What do you see around you?
* What do you hear?

* What sensations do you feel in your body?
* What emotions arise?
* Are there tastes or smells?

Live in this completed reality for full 60 seconds

Visualization Segment 3 Energetic Broadcast (30 seconds):

Feel your coherent heart sending this reality's frequency

- See the frequency radiating from your heart into the world

Know that this frequency is attracting your manifestation

STEP 5: INSPIRED ACTION RECEPTION (2 minutes)

Maintain hand on heart

Pose the question: "What is one inspired action I can take today toward this manifestation?"

Don't force an answer—remain open

Accept whatever guidance or idea emerges

Commit to taking this action today

This bridges the gap between internal

alignment and external action

STEP 6: ENERGY FIELD ACTIVATION (2 minutes)

Part A: Personal Field Strengthening (1 minute)

Visualize brilliant white light at your heart center

See this light expanding to fill your entire body

Feel it extending 3 feet in all directions around your body

Set the intention: "This field attracts opportunity and repels obstacles"

Feel your entire being surrounded in protective, magnetic light

Part B: Daily Frequency Anchor (1 minute)

Choose a physical gesture you can repeat throughout the day (hand on heart, specific hand position, power pose, etc.)

Practice your gesture 3 times while feeling the powerful frequency

Choose a simple phrase for busy moments

(Example: "I am aligned," "I am worthy," "I am abundant")

Say phrase 3 times while feeling the established frequency

Know that you can access this frequency anytime by repeating your gesture

STEP 7: INTEGRATION & TRANSITION (1 minute)

Stand up slowly

Feel powerful energy field moving with you

Final statement: "I go forth today as a conscious creator,

aligned with my highest manifestation. I am the frequency of [intention]."

Move intentionally into your morning with presence

Total Time: 15 minutes

Manifestation Impact: POWERFUL daily recalibration, consistent results within 2 3 weeks

EXTENDED VERSION (30 minutes)

For Deeper Work

Use on weekends, during manifestation focused periods, or when you want more profound recalibration. This version adds deeper layers and extended focus periods.

STEP 1: SACRED SPACE & GROUNDING (2 minutes)

Create more elaborate sacred space if desired

Light candle, burn incense or essential oil

Sit on cushion with feet grounded or in chair with feet flat

Ring bell or chimes 3 times to mark the beginning

STEP 2: EXTENDED PRANAYAMA (5 minutes)

Technique 1: Kapalbhati (Skull Shining Breath) 2 minutes

Rapid, forceful exhalation through nose, passive inhalation

Perform 20 30 quick exhalations, then pause and breathe normally

Repeat 3 times total

Purpose: Energize, clear mental

cobwebs, prepare nervous system

Technique 2: Nadi Shodhana (Alternate Nostril) 2 minutes

More extended version with longer counts

Close right, inhale left (8 counts)

Close left, exhale right (8 counts)

Inhale right (8 counts), close right, exhale left (8 counts)

Continue 20 full cycles

Purpose: Perfect balance, activate subtle energy channels

Technique 3: Ujjayi (Ocean Breath) 1 minute

Slight throat constriction, creating ocean wave sound

Inhale fully, exhale with gentle throat constriction for 8 counts

Continue for 1 minute

Purpose: Create soothing, meditative state, focus mind

STEP 3: EXTENDED CHAKRA ACTIVATION (6 minutes)

For each chakra, hold the sound for

20 30 seconds and add visualization:

ROOT CHAKRA LAM (1 minute)

Sound vibration at base of spine

Visualize red light spiraling at base

Feel stability, grounding, security

Say: "I am grounded in my power"

SACRAL CHAKRA VAM (1 minute)

Sound vibration in lower abdomen

Visualize orange light spiraling

Feel creative energy flowing

Say: "I create with joy and flow"

SOLAR PLEXUS RAM (1 minute)

Sound vibration in upper abdomen

Visualize yellow light spiraling

Feel personal power activating

Say: "I am powerful and capable"

HEART CHAKRA YAM (1 minute)

Sound vibration in chest

Visualize green/pink light spiraling

Feel love and connection expanding

Say: "I love and am loved"

THROAT CHAKRA HAM (1 minute)

Sound vibration in throat

Visualize blue light spiraling

Feel your truth expressing clearly

Say: "I speak my truth with power"

THIRD EYE OM (1 minute)

Sound vibration in forehead

Visualize indigo light spiraling

Feel intuition awakening

Say: "I see clearly with wisdom"

CROWN CHAKRA Pure Silence (variable)

Allow the Om to continue softly into silence

Feel connection to divine/
higher consciousness

Simply exist in the infinity of space

STEP 4: EXTENDED MANTRA CHANTING (6 minutes)

Primary Manifestation Mantra   3 minutes

Chant chosen manifestation mantra

Chant slowly and deliberately, feeling each sound

Approximately 1 2 chants per breath cycle

Let the mantra penetrate deeply into your being

Secondary Alignment Mantra   2 minutes

Choose a mantra that reinforces your daily intention

Example for alignment: "So Hum" (I am that)

Example for divine support: "Om Namah Shivaya"

Example for grace: "Om Gam Ganapataye Namaha"

Silence   1 minute

Let the mantras settle

Sit in the vibration they've created

Allow the sound to continue internally

STEP 5: EXTENDED VISUALIZATION (7 minutes)

Part A: Body Blessing (2 minutes)

Send appreciation to each major body system

Bless your nervous system for its wisdom

Bless your heart for its love

Bless your mind for its clarity

Bless your digestive system for its transformation

Bless your immune system for its protection

Bless your lymph system for its flow

Part B: Ancestral Gratitude (2 minutes)

Connect with your family lineage

Thank your ancestors for their blessings

Acknowledge the strengths they passed to you

Ask for their continued support in your manifestations

Visualize their loving energy supporting your goals

Part C: Future Self Connection (2 minutes)

Project yourself 1 2 years into the future

Connect with the version of yourself that HAS manifested your goals

Feel what that version feels like

Ask that future self: "What was the most important frequency shift I needed to make?"

Listen for the answer

Thank your future self for showing you the way

Part D: Primary Manifestation Visualization (1 minute)

Final confirmation of your manifestation as already complete

Feel the emotions of success, gratitude, joy

Affirm: "This is mine. This is done. This is flowing to me now."

STEP 6: ENERGY FIELD EXPANSION (3 minutes)

Part A: Personal Field (1 minute)

Brilliant white light at heart expanding through body

Extending 3 feet in all directions

Protection and magnetic attraction activated

Part B: Heart Field Expansion (1 minute)

Heart light expanding to fill entire home/space

Creating sanctuary frequency in your environment

Anyone who enters experiences your high vibration

Part C: Universal Connection (1 minute)

Heart light expanding to fill your city

Extending to fill your country

Extending to encompass the entire planet

Know that your coherent frequency supports the collective

STEP 7: INTEGRATION & DAILY ANCHORING (1 minute)

Slowly return awareness to body

Feel the complete integration of all practices

Choose gesture and phrase for daily access

Stand and move with intention

Carry this frequency consciously throughout your day

Total Time: 30 minutes

Manifestation Impact: PROFOUND recalibration, noticeable results within 1 week

SUCCESS INDICATORS

Physical Signs (What You'll Feel):

Deep relaxation despite being alert

Warm tingling in hands and feet

Chest opening and expansion

Breathing becoming slower and deeper naturally

Muscles releasing tension

Body feeling lighter and more energized

Heart rate slowing to calm rhythm

Possible spontaneous movements or stretches

Mental Signs (What You'll Think):

Mental chatter significantly quieting

Clarity about what you need to do today

Shift from stress mindset to abundance mindset

Creative ideas emerging naturally

Sense of calm confidence

Mental obstacles seeming smaller or dissolving

Increased focus and concentration

Worries fading into background

Emotional Signs (What You'll Feel):

Deep peace and contentment

Genuine gratitude for your life

Excitement about possibilities

Joy and lightness

Self compassion and self love

Confidence in your abilities

Connection to something larger than yourself

Emotional resilience

Manifestation Signs (Evidence It's Working):

During week 1: Increased clarity about goals

During week 2: Opportunities appearing that align with frequency

During week 3: People responding differently to you

During week 4: Tangible progress

toward manifestations

Ongoing: Synchronicities increasing, ease of flow, reduced resistance

Energy Field Signs (What Your Aura Looks Like):

Bright, coherent, expanding

Magnetic—people naturally drawn to you

Protective—unwanted energies bounce away

Stable—grounded yet elevated

Coherent—no fragmentation or chaos

Timeline:

First session: Immediate shift in mental and emotional state

Consistent daily practice (1 2 weeks): Noticeable ease in life

One month: Significant manifestation acceleration

Three months: Complete nervous system recalibration, new baseline established

TROUBLESHOOTING

If "I don't have time for the full practice":

Use Quick Version (10 min) on busy days

Even Quick Version maintains core benefits

Find time on weekends for Extended Version

Consistency at 10 min sporadic at 30 min

If "I fall asleep during the practice":

This is fine—your subconscious is receiving the imprinting

If you want to stay awake, try practicing before breakfast

Sit upright in a chair instead of lying down

Cold water splash before practice helps with alertness

If "I don't feel anything special":

Give it 7 consecutive days before assessing

Some people experience subtle shifts they don't immediately recognize

Track physical signs (stress, sleep, energy) as evidence

Remember: the practice works whether you feel it or not

If "I keep getting distracted":

This is completely normal—mind

wandering is the practice

Each time you notice and return, you're strengthening focus

Use a mala (prayer beads) to give hands something to do

Simplify the practice if too many components

If "Nothing seems to be manifesting":

This practice lays the foundation

Combine with other practices from this book

Take at least one action daily aligned with your intention

Track 30 40 days before assessing results

If "I feel emotional or want to cry":

Perfect—the practice is releasing blocked emotion

Allow the emotions to flow

This emotional clearing accelerates manifestation

It may take several days for this to complete

If "My mind races with a to do list":

Write down your tasks before practice

Knowing they're written allows

your mind to release them

Or do the practice slightly earlier before thinking about the day

If "I'm too busy in the morning":

Wake 10 15 minutes earlier—this time creates more than it takes

Or split into 2 shorter sessions (morning evening)

Or move to evening if morning is impossible

Consistency matters more than timing

INTEGRATION

Use Quick Version (10 min) on:
Busy weekday mornings

Use Standard Version (15 min) on: Most mornings for consistent results

Use Extended Version (30 min) on: Weekends or during focused manifestation periods

Combine this practice WITH:

Evening Energy Shower (Chapter 10) for complete daily cycle

Specific mantras for targeted

manifestation work

Physical movement to ground the energy into action

Journaling to clarify insights that emerge

Do NOT use as replacement for:

Meditation (though this contains meditative elements)

Physical exercise (though this provides some energetic activation)

Taking inspired action (practice action results)

Best Timing:

Ideal: First thing upon waking (frequencies are cleanest)

Acceptable: Before breakfast

Less ideal but workable: Any morning time before day gets hectic

Emergency: Split into two 7 8 minute sessions

Timeline to Results:

Days 1 3: Mental clarity, slight mood improvement

Days 4 7: Noticeable calm, opportunities emerging

Days 8 21: Significant manifestation acceleration, synchronicities increasing

Days 22 40: Complete nervous system recalibration, manifestation becomes natural

40 days: New baseline established, benefits automatic

Scientific Validation:

HeartMath research confirms coherence effects within 2 3 minutes

Brainwave entrainment literature shows state shifts within 10 15 minutes

Breath work research documents nervous system regulation within 5 minutes

Mantra studies show measurable frequency shifts within 10 minutes

Combined, these create a POWERFUL cascade of effects

Daily Commitment:

This single practice, done consistently, can more than double your manifestation results. It's the highest ROI practice in this entire book.

Seasonal Variations

Spring (March May): Face east, add green visualiza tion, focus on new projects
Summer (June August): Include golden solar energy, extended success mantras
Autumn (September November): Begin with gratitude for received manifestations
Winter (December February): More meditation time, contemplative mantras, spiritual focus

◆ ◆ ◆

CHAPTER 9: MICRO-RESETS DURING THE DAY

3-Breath Reset

You started your day at a high frequency.

Then one thing happened, and it all collapsed.

What if you could reset in 60 seconds—no matter where you are, no matter what happened?

This chapter gives you 7 tools to recalibrate your entire frequency

instantly throughout your day.

This is where the 3-Breath Reset becomes invaluable. This simple yet powerful technique allows you to shift your vibrational state in under a minute, no matter where you are or what's happening around you. It's your emergency frequency modulator—a way to quickly return to alignment when life knocks you off center.

Why Three Breaths?

Three breaths is the sweet spot for a micro-reset. One breath isn't quite enough to break an established pattern, while more than three often requires a dedicated break that isn't always practical. Three conscious breaths:

Takes approximately 30 seconds—short enough to do anywhere

Is long enough to interrupt negative patterns

Creates a complete energetic circuit

Can be done discreetly in virtually any situation

Is easy to remember even when stressed

The power of this technique lies not just in the breathing itself but in the specific focus and intention you bring to each of the three breaths. Each breath serves a distinct purpose in the reset process.

The 3-Breath Reset Technique

When you notice your frequency dropping—when you feel stressed, anxious, angry, or otherwise misaligned with your manifestation state—implement this technique immediately:

Breath One: Release

The first breath is all about letting go of the current discordant frequency.

Inhale deeply through your nose, filling first your belly, then your chest

As you inhale, mentally gather all the tension, stress, or negative emotion you're experiencing

Hold the breath briefly at the top

Exhale forcefully through your mouth with a sighing sound

Visualize the discordant energy leaving your body completely with this exhale

Mentally say "Release" as you exhale

This first breath creates space by clearing the dominant lower frequency, making room for the new vibration you want to establish.

Breath Two: Reset

The second breath neutralizes your field and returns you to a centered state.

Inhale slowly and deeply through your nose

As you inhale, visualize pure white light entering your body

Direct this light to the center of your chest, creating a glowing sphere in your heart center

Hold the breath briefly, feeling this light expanding within you

Exhale slowly through your nose, allowing the light to spread throughout your body

Mentally say "Reset" as you exhale

This second breath establishes a neutral baseline—like rebooting a computer that's running too many programs and becoming sluggish.

Breath Three: Realign

The third breath activates your chosen frequency and realigns you with your manifestation intention.

Inhale deeply through your nose

As you inhale, visualize golden light entering from above

See and feel this light carrying the specific frequency you wish to embody (confidence, peace, abundance, etc.)

Hold the breath briefly, allowing this frequency to saturate every cell

Exhale slowly through your nose, extending this frequency beyond your body into your energy field

Mentally say "Realign" as you exhale

This third breath actively establishes the new frequency you choose to broadcast, reconnecting you with your manifestation intention.

Amplifying the Effect

To enhance the power of the 3-Breath Reset, add these elements when possible:

Physical Anchors

Pair the technique with subtle physical gestures that can be done anywhere:

Touch your thumb and middle finger together during the process

Place one hand over your heart center

Straighten your posture and roll your shoulders back

Slightly raise your gaze upward on the third breath

These physical anchors help integrate the frequency shift throughout your body and create a conditioned response that makes future resets more effective.

Frequency Words

Select a specific word that represents the frequency you want to embody, and focus on this word during the third breath. Examples:

"Clarity" when feeling confused

"Peace" when feeling anxious

"Power" when feeling weak

"Abundance" when feeling lack

This word becomes a vibrational shortcut that your energy field learns to respond to with practice.

Sensory Enhancement

Engage multiple senses to strengthen the reset:

Feel the temperature and texture of the air as you breathe

Listen to the sound of your breath

Visualize the colors associated with your desired frequency

Notice the physical sensations of energy moving in your body

The more senses you engage, the more complete the reset becomes.

Strategic Implementation

To get the most benefit from the 3-Breath Reset, use it:

Preventatively

Don't wait until you're fully triggered into a low frequency.

Learn to recognize the early warning signs that your vibration is beginning to drop, and implement the reset at the first indication. These signs might include:

Slight tension in your shoulders or jaw

A subtle shift in your thought patterns toward worry or criticism

A feeling of energy draining from your body

Decreased enthusiasm or increased irritability

Transitionally

Use the reset when moving between different environments or activities:

Before entering a meeting

After a difficult conversation

When arriving home from work

Before beginning creative or manifestation work

When switching between projects

These transition points are natural opportunities to reset your frequency before it becomes compromised.

Rhythmically

Schedule regular resets throughout your day:

Once every hour

Before meals

At the top of each hour

Before making important decisions

After exposure to news or social media

These rhythmic resets maintain your vibrational baseline even when you don't feel particularly off-center.

Environmentally

Certain locations tend to affect your frequency in predictable ways. Use the reset when entering:

Crowded public spaces

Hospitals or medical facilities

High-stress work environments

Places with difficult emotional associations

Anywhere with chaotic or dense energy

Real-World Applications

The 3-Breath Reset can be adapted for specific situations:

In Meetings

When you feel your energy dropping during a meeting, excuse yourself to use the restroom if possible. If not, simply sit up straighter and perform the reset subtly while others are speaking.

During Conflicts

When tension arises in a relationship, create a brief pause by saying, "Let me think about that for a moment." Use this pause to perform the reset before responding.

While Commuting

Traffic and public transportation can significantly lower your frequency. Use red lights or stops as opportunities to perform the reset.

Before Important Communications

Before sending important emails, making phone calls, or having significant conversations, take 30 seconds for the 3-Breath Reset to ensure you're broadcasting the frequency that supports your intentions.

With practice, this technique becomes second nature—a reflexive response to frequency disruptions that allows you to maintain vibrational consistency throughout your day. This consistency is crucial for manifestation, as it's your predominant frequency, not just your peak moments, that determines what you attract into your experience.

The 3-Breath Reset gives you a portable, practical tool to manage your vibrational state in real-time, ensuring that the frequency you established in your morning practice remains your dominant broadcast throughout the day.

1-Song Vibration Lift

Music isn't just entertainment—it's one of the most powerful and immediate frequency modulators available. The right song at the right moment can completely transform your vibrational state in just a few minutes, making it an invaluable tool for manifestation maintenance throughout your day.

This technique harnesses the vibrational power of music for quick, effective frequency shifts when you need them most. Unlike passive music listening, the 1-Song Vibration Lift is an active practice that maximizes music's effect on your energy field.

The Science Behind Musical Frequency Shifts

Music affects your vibrational state through multiple mechanisms:

Entrainment: Your brainwaves naturally synchronize with the rhythm and tempo of music, allowing you to shift states by selecting music with specific beats per minute (BPM).

Emotional Resonance: Music directly activates the limbic system—your emotional brain—triggering the release of neurotransmitters and hormones that alter your biochemical state.

Memory Anchoring: Songs become associated with emotional states and experiences, allowing them to reactivate those states when heard again.

Harmonic Influence: Different musical scales and harmonies create distinct energetic effects on your nervous system and energy field.

Lyrical Programming: Song lyrics can function as affirmations or pattern interrupts for your conscious and subconscious mind.

By combining these effects with conscious intention, you can use a single song as a complete vibrational reset tool.

Creating Your Vibration Lift Playlist

The first step is creating a specialized playlist of songs specifically selected for their frequency-shifting effects. This isn't just a collection of songs you enjoy—it's a strategic toolkit for vibrational management.

Create categories in your playlist for different frequency needs:

Grounding Songs (For when you feel scattered or anxious)

Look for songs with:

Steady, slower tempos (60-80 BPM)

Earth-toned instruments (drums, bass, cello)

Simple, repetitive structures

Lyrics about presence, stability, or strength

Examples: "Teardrop" by Massive Attack, "Breathe Me" by Sia, "Holocene" by Bon Iver

Energizing Songs (For when you feel depleted or low)

Look for songs with:

Upbeat tempos (100-130 BPM)

Major keys and uplifting harmonies

Building dynamics that increase throughout the song

Lyrics about power, possibility, or overcoming

Examples: "Freedom" by Beyoncé, "High Hopes" by Panic! At The Disco, "Unstoppable" by Sia

Heart-Opening Songs (For when you feel closed or disconnected)

Look for songs with:

Flowing melodies with emotional resonance

String instruments or piano

Vulnerable, authentic vocal delivery

Lyrics about love, connection, or compassion

Examples: "All of Me" by John Legend, "Fix You" by Coldplay, "River Flows in You" by Yiruma

Manifestation Amplifier Songs (For directly boosting manifestation power)

Look for songs with:

Triumphant, expansive qualities

Lyrics that align with abundance, success, or dreams coming true

Powerful climaxes or chorus sections

Personal significance related to your specific goals

Examples: "Hall of Fame" by The Script, "This Is Me" from The Greatest Showman, "Roar" by Katy Perry

For each category, select 3-5 songs that reliably create the desired state for you personally. What works for someone else might not work for you—the key is your personal resonance with the music.

The 1-Song Vibration Lift Technique

When you need a vibrational reset during your day, follow these steps:

Frequency Assessment (10 seconds)

Briefly check in with your current vibrational state

Identify what frequency you need to shift toward

Select the appropriate song from your playlist

Environment Optimization (20 seconds)

If possible, find a private space

Put on headphones for full immersion (much more effective than speakers)

Adjust volume to an optimal level (loud enough to be immersive but not uncomfortable)

Close your eyes if the situation allows

Intentional Listening (Duration of song)

Begin with a clear intention: "This song is shifting my frequency to [desired state]"

Focus completely on the music, avoiding distractions

Breathe in rhythm with the music when possible

Allow the song to move through your entire body, not just your ears

Physical Engagement (Throughout song)

Move your body in some way, even if subtle

Tap your fingers or feet to the rhythm

If privacy allows, dance, sway, or move freely

Use hand gestures to "conduct" the emotional peaks of the music

Vocal Participation (When appropriate)

Hum, sing along, or vocalize with the music

Pay attention to how the vibration feels in your chest and throat

If you know the lyrics and they're aligned with your intention, sing them with conviction

If the lyrics don't serve your intention, focus on the melody instead

Frequency Lock-In (15 seconds after song ends)

Keep your eyes closed for a moment after the song ends

Take one deep breath to integrate the new frequency

Place your hand on your heart and silently affirm: "This frequency is now mine"

Open your eyes and carry this state into your next activity

Maximizing the Effect

To get the most from this technique:

Pre-condition Your Songs

You can increase a song's effectiveness by intentionally using it during peak states:

When you naturally feel the desired state, play your selected song

Fully associate the song with this peak state through repeated pairing

Over time, the song becomes a reliable trigger for that state

Layer with Other Techniques

Combine the 1-Song Vibration Lift with:

The 3-Breath Reset before and after the song

Visualization during instrumental sections

Affirmations that align with the music's energy

Movement or stretching to amplify the physical component

Strategic Timing

Plan your 1-Song resets at key points in your day:

Mid-morning energy dip (around 10:30 am)

After lunch to prevent afternoon sluggishness

Before important meetings or calls

After exposure to lower frequencies (difficult people, news, etc.)

As a transition between work and personal time

Progressive Lifting

For deeper frequency shifts, use a sequence of 2-3 songs that progressively move you toward your desired state:

Start with a song that meets you where you are

Move to a transitional song that begins the shift

Finish with a song that embodies your target frequency

Real-World Applications

This technique can be adapted for various situations:

At Work

Keep a pair of headphones at your desk for quick resets between tasks. Even stepping away for a 3-4 minute musical reset can dramatically change your productivity and manifestation alignment.

While Commuting

Transform otherwise wasted time into vibrational management by using your commute for intentional music medicine. Create specific playlists for your morning and evening commutes to bookend your workday with aligned frequencies.

During Exercise

Exercise already changes your biochemistry—pairing it with intentional music amplifies the frequency benefits. Create workout playlists specifically designed for vibrational lifting rather than just motivation.

Before Sleep

Use a specially selected song to shift your frequency before sleep, programming your subconscious for aligned manifestation work during dream time.

The 1-Song Vibration Lift gives you a portable frequency modulation tool that works in virtually any situation. By creating a personalized musical pharmacy and using it with intention, you gain the ability to shift your vibrational state on demand throughout your day.

Music isn't just something you listen to—it's something you can use as a conscious creator to maintain the frequency of what you wish to manifest.

Movement Breaks

Your body isn't just a vehicle for your consciousness—it's an integral part of your vibrational system. The way you move, hold, and use your body directly affects your frequency and, by extension, your manifestation abilities.

When your energy becomes stagnant or your frequency drops during the day, strategic movement can create immediate shifts more effectively than mental techniques alone. These physical micro-resets bypass the thinking mind and work directly with your energetic body to restore optimal vibrational states.

Here are three powerful movement practices you can use anywhere to quickly shift your energy and raise your vibration:

The Shake Technique: Releasing Stagnant Energy

Shaking is one of the most primal and effective ways to release blocked energy and reset your nervous system. Animals instinctively shake after stressful encounters to discharge tension—humans can use this same mechanism intentionally.

The Basic Technique (1-2 minutes):

Stand with your feet shoulder-width apart, knees slightly bent

Begin by gently shaking your hands as if drying them

Gradually increase the intensity and let the movement spread up your arms

Allow your shoulders to join the shaking motion

Let the vibration move down through your torso, hips, and legs

Continue for 1-2 minutes, allowing your whole body to shake

Make sound if possible—sighing, humming, or vocal tones enhance the release

After shaking, stand still for 30 seconds and notice the energy flowing through your body

Vibrational Benefits:

Breaks up energetic blockages in your physical and subtle bodies

Resets your nervous system from "fight or flight" to "rest and digest"

Releases emotional energy stored in muscle tissue

Creates space for new, higher frequencies to enter your field

Discrete Variations:

When a full-body shake isn't appropriate, try these subtle alternatives:

Under a desk, shake just your legs for 30 seconds

In a bathroom stall, shake your hands vigorously for 20 seconds

While walking between meetings, shake out your hands and arms

Sitting in your car, vibrate your lips with a "brrr" sound for 10 seconds

Sun Gazing: Absorbing High-Frequency Light

The sun is the most powerful source of vibrational energy in our solar system. Brief, safe sun gazing can rapidly elevate your frequency by connecting you directly to this primal energy source.

The Basic Technique (1-3 minutes):

This practice is ONLY safe during the first 30 minutes after sunrise or the last 30 minutes before sunset when the UV index is low

Stand barefoot on natural ground if possible

Remove glasses or contacts if you wear them

Gaze softly at the sun for 10-30 seconds (start with less time and gradually build)

Close your eyes for 10 seconds and observe the after-image

Repeat this cycle 3-5 times

After completing the cycles, place your palms over your closed eyes for 30 seconds

Finish by placing your hands on your heart and feeling the solar energy circulating through your body

Vibrational Benefits:

Directly absorbs photonic energy that raises cellular vibration

Stimulates the pineal gland, associated with higher consciousness

Activates the visual cortex in ways that affect brainwave patterns

Creates a direct energetic connection to the primary source of life energy in our system

Discrete Variations:

When full sun gazing isn't possible:

Gaze at the sky in the direction of the sun (not directly at it)

Stand by a window where sunlight is entering and absorb it through closed eyelids

Place your hands in direct sunlight and imagine absorbing the energy through your palms

Visualize the sun above your head, sending golden light into your body

Nature Walking: Harmonic Resonance Reset

Natural environments vibrate at frequencies that automatically harmonize human energy fields. A brief nature walk creates a vibrational reset through multiple mechanisms.

The Basic Technique (5-10 minutes):

Find any natural setting—a park, garden, or even a single tree

Begin walking at a comfortable pace

Synchronize your breath with your steps (e.g., 3 steps inhale, 3 steps exhale)

Touch natural elements as you pass—leaves, bark, stones

With each breath, imagine exchanging energy with the natural world

Pause occasionally to stand still and feel the subtle energy of your surroundings

Before returning to your regular activities, place your hand on a tree or plant and mentally thank it for the energy exchange

Vibrational Benefits:

Natural settings emit negative ions that counteract the positive ions from electronic devices

The fractal patterns in nature automatically organize chaotic brain activity

Earth's electromagnetic field helps reset your personal field to its natural state

The diverse frequencies of a natural ecosystem create beneficial entrainment

Discrete Variations:

When a full nature walk isn't possible:

Step outside your workplace for 60 seconds of sky gazing

Keep a small plant on your desk and take "micro-breaks" to connect with it

Hold a natural object (stone, wood, leaf) while taking three deep breaths

Look at images of nature scenes for 30 seconds while imagining being there

Implementing Movement Breaks in Your Day

To make these practices part of your regular frequency maintenance:

Schedule Triggers

Rather than waiting until your energy crashes, establish regular movement break triggers:

After 60 minutes of sitting

Between different types of tasks

Following challenging interactions

Before important creative or manifestation work

When transitioning between environments

Environment Preparation

Create supportive conditions for quick movement breaks:

Keep comfortable shoes under your desk for quick outdoor access

Identify private spaces where you can shake without observation

Map nearby natural areas for brief visits during your day

Note east/west facing windows for morning/evening sun connection

Accountability Systems

Make movement breaks non-negotiable:

Set alarms or calendar reminders

Partner with a colleague for mutual encouragement

Track your breaks and their effects in a simple log

Connect breaks to existing habits (e.g., after every bathroom visit)

Mindful Integration

Maximize the benefit of each movement break:

Set a clear intention before beginning

Remain fully present during the movement

Notice the before/after difference in your state

Carry the shifted energy into your next activity

The Vibrational Mechanics of Movement

These practices work because movement affects your frequency through multiple pathways:

Physical Pathway:

Movement increases circulation, delivering more oxygen to your cells and enhancing their energetic function. It also releases tension in muscle tissue where emotional energy is stored.

Electrical Pathway:

Movement changes the electrical activity in your nervous system, affecting brainwave patterns and heart rhythm coherence—both key factors in your vibrational broadcast.

Chemical Pathway:

Movement triggers the release of neurotransmitters and hormones that directly affect your emotional state and energy level, shifting you toward higher vibrational frequencies.

Energetic Pathway:

Movement stimulates the flow of subtle energy (prana/chi) through your meridian system and chakras, clearing blockages and enhancing overall energy circulation.

By incorporating these movement breaks into your day, you create regular opportunities to reset and elevate your vibrational state regardless of external circumstances. This consistent frequency management ensures that your manifestation efforts aren't undermined by prolonged periods of lower vibration.

Your body isn't separate from your manifestation work—it's an essential instrument for maintaining the frequency of what you wish to create.

Advanced Vibrational Reset Techniques

Your body is a sophisticated bioelectrical instrument that accumulates energetic debris throughout the day. These movement breaks serve as vibrational "system resets" that clear stagnant energy and realign your frequency with your manifestation goals.

The Science of Movement-Based Energy Clearing

How Energy Gets Stuck:

- Postural tension creates electromagnetic field restrictions
- Mental stress and anxiety create coherent energy blockages
- Unexpressed emotions lodge in muscle tissue and energy field
- Environmental frequency pollution disrupts natural biorhythm
- Unconsciously absorbing others' stress and emotional states

Signs You Need a Vibrational Reset:

- Mental fog or decision-making difficulty
- Physical tension in shoulders, neck, or jaw
- Feeling "heavy" or sluggish despite adequate rest
- Irritability or emotional reactivity
- Loss of clarity about manifestation goals
- Decreased intuitive sensitivity

The Shake Technique: Full Body Vibrational Reset

Basic Shake Protocol (3-5 minutes)

Preparation Phase (30 seconds):

1. Stand with feet shoulder-width apart, knees slightly bent
2. Intention: "I release all energy that doesn't serve my highest manifestation"

3. Take three deep breaths, feeling your body's current energetic state
4. Give your body permission to move however it needs for clearing

Level 1: Gentle Awakening (60 seconds)

- Begin by shaking hands loosely at wrists
- Gradually include forearms, then entire arms
- Breathe naturally, don't control the movement
- Feel tingling or warmth as energy begins to move
- Visualize stagnant energy being shaken loose like dust from fabric

Level 2: Full Body Integration (90 seconds)

- Add shoulder rolling and shaking
- Let shaking movement ripple through chest and belly
- Include hip movement and pelvic circles
- Allow shaking to travel down through legs to feet
- Let natural sounds emerge (sighs, groans, vibrations)
- See your entire energy field becoming fluid and dynamic

Level 3: Deep Discharge (60 seconds)

- Increase shaking vigor as feels natural
- Full-body shaking with jumping or bouncing
- Include facial shaking and jaw movement
- Add vocal toning or shaking sounds
- Let movement gradually slow naturally
- See all stuck energy flying off your body like water drops

Integration Phase (30 seconds):

- Stand quietly feeling the energy moving through your body
- Notice areas that feel different, more relaxed, or energized
- Thank your body for its wisdom in releasing what didn't serve
- Set intention to maintain this clear energy state

Advanced Shake Variations

Emotional Shake (when feeling emotionally stuck):

- Focus shaking on chest, shoulders, and throat
- Use forceful exhales to discharge emotional energy
- Allow crying, laughing, or vocal release as needed
- Continue until emotional charge feels cleared (3-10 minutes)

Mental Clarity Shake (when mind feels foggy):

- Emphasize head, neck, and shoulder shaking
- Use rapid breath patterns (like Kapalabhati) during shaking
- Include head shaking (gently) and jaw loosening
- Continue 2-3 minutes until mental clarity returns

Manifestation Power Shake (before important manifestation work):

- Full body with emphasis on core and hands
- Include power gestures and strong grounding through feet
- Visualize yourself as powerful electromagnetic field broadcasting desires
- State: "I activate my manifestation power"

Advanced Movement Breaks

The Spinal Wave Reset (2-3 minutes)

- Stand with feet hip-width apart, arms at sides
- Begin subtle forward and backward rocking motion
- Let movement create gentle wave-like motion up and down spine
- Coordinate breath: inhale as wave moves up, exhale as moves down
- Gradually increase amplitude and fluidity
- End with gentle side-to-side spinal movement

The Heart Opening Power Break (2-4 minutes)

- Large backward arm circles for 30 seconds
- Interlace fingers behind back, lift chest, hold 30 seconds
- Press forearms against doorframe, lean forward 30 seconds
- Place hands on heart, breathe deeply into chest area
- Affirmation: "My heart is open to receiving all my manifestations"

The Grounding Stomp Series (1-2 minutes)

- Stomp feet alternately, feeling connection with Earth
- Stomp with both feet simultaneously
- Include arm movements that mirror grounding energy
- Use forceful exhales with each stomp
- Intention: "I ground myself in the supportive energy of Earth"
- Stand quietly feeling solid and stable

Workplace-Appropriate Movement Breaks

The Desk Warrior Reset (90 seconds)

Seated Variations (45 seconds):

- 10 backward shoulder rolls

- Gentle neck circles and side stretches
- Seated spinal twists holding chair back
- Circle both feet to activate leg energy flow
- Shake hands and wrists under desk

Standing Variations (45 seconds):

- Rise onto toes 10-15 times
- Subtle hip circles to release lower back
- 5 deep breaths with arm raises
- Pull shoulder blades together, release
- Silent appreciation for body's service

The Bathroom Break Reset (60 seconds)

- 30-second full-body shake in private bathroom stall
- Massage temples, jaw, and forehead
- 5 deep breaths with visualization of clear energy
- Reset intention for remainder of day
- Exit feeling refreshed and realigned

Daily Movement Break Schedule

Optimal Timing:

- **Morning Activation**: Full 5-minute reset upon waking
- **Mid-Morning Reset** (10 AM): 2-3 minute shake after first work hours
- **Lunch Break Reset**: Complete 5-minute sequence if possible
- **Afternoon Prevention** (3-4 PM): 90-second desk reset
- **Evening Transition**: Full shake to discharge workday stress
- **Pre-Sleep Reset**: Gentle movements to release day's tensions

Integration with Other Practices

Movement + Sound:

- Practice Om chanting during movement breaks
- Coordinate breath and sound with shaking movements
- Use movement to physically integrate mantra vibrations

Movement + Frequency Audio:

- Play clearing frequencies (396 Hz, 741 Hz) during movement
- Let frequency guide natural rhythm and intensity
- Combine workplace movement with earphones playing frequencies

Pre/Post Manifestation:

- Always do movement break before visualization work
- Clear energy state enhances manifestation effectiveness
- Gentle movement after intensive visualization helps integrate shifts

CHAPTER 10: EVENING FREQUENCY RESET

Energy Shower

After a full day of navigating various environments, interactions, and challenges, your energy field naturally accumulates debris—much like your physical body collects dust and sweat. This energetic residue, if not cleared, affects your sleep quality, subconscious programming, and next-day manifestation abilities.

The Energy Shower technique provides a complete energetic cleansing that's as essential to your vibrational hygiene as physical bathing is to your bodily hygiene. This practice creates energetic space for renewal and prevents carrying low vibrations into your sleep state, where they can affect your subconscious manifestation work.

The Energetic Accumulation Problem

Throughout your day, your energy field collects various types of residue:

Emotional residue from your own reactions and others' emotional broadcasts

Thought-form attachments from repetitive or intense thinking patterns

Energetic cords created during interactions with others

Environmental frequencies absorbed from spaces you've occupied

Electromagnetic pollution from devices and artificial fields

This accumulated energy affects you in several ways:

It creates a filter through which you perceive reality

It influences your emotional baseline and reactivity

It can block the reception of intuitive guidance

It interferes with your clear vibrational broadcast for manifestation

Just as you wouldn't go to bed covered in the day's physical dirt, you shouldn't sleep wrapped in the day's energetic debris.

The Complete Energy Shower Technique (7-10 minutes)

This comprehensive practice cleanses your entire energy field, preparing you for restorative sleep and effective subconscious manifestation work.

Preparation:

Choose a time when you won't be disturbed, ideally at least 30 minutes before sleep

Turn off electronic devices or put them in airplane mode

Dim the lights to create a calming environment

Optional: Light a candle to represent transformation

Stand or sit comfortably with your spine straight

Step 1: Awareness Scan (1-2 minutes)

Begin by bringing conscious awareness to your current energetic state:

Close your eyes and take three deep breaths

Starting at the top of your head, slowly scan down through your body

Notice areas that feel heavy, tight, or uncomfortable

Pay attention to any emotions still lingering from the day

Observe your overall energy level and quality

Set a clear intention: "I now cleanse my energy field of all that doesn't serve my highest good and manifestation goals."

Step 2: Breath Preparation (1 minute)

Activate your energy system with conscious breathing:

Inhale deeply through your nose for a count of 4

Hold your breath for a count of 4

Exhale completely through your mouth for a count of 6

Repeat this cycle 4 times

On the final exhale, imagine releasing any obvious tension or stress

Step 3: The Waterfall Visualization (2-3 minutes)

Now begin the primary cleansing visualization:

Imagine a waterfall of brilliant white or golden light forming above your head

See this light as pure universal energy—cleansing, revitalizing, and protective

Allow this waterfall to begin flowing down over your head and face

Feel it washing away all stress, worry, and negative thought patterns

Let it continue flowing down your neck and shoulders, dissolving tension

See it moving down your arms to your fingertips, flushing out any energy that isn't yours

Feel it flowing down your torso, cleansing your heart of emotional residue

Allow it to continue down your legs to your feet, grounding and clearing your entire system

Visualize this energy continuing down into the earth, carrying away all that you've released

See your entire body and energy field glowing, clean and clear

Step 4: Cord Cutting (1-2 minutes)

Address specific energetic connections that may have formed during the day:

Bring to mind the people you interacted with today

For each significant interaction, check if you feel any energetic cords or attachments

These might feel like pulls, hooks, or threads connecting from your body to elsewhere

For each cord you sense:

Thank it for what it has shown you

State firmly: "I now release this connection with love"

Visualize the cord dissolving, being cut, or gently unplugging

Affirm: "I am complete unto myself. All connections that serve my highest good remain in perfect balance."

Step 5: Boundary Restoration (1 minute)

Reestablish your energetic boundaries:

Visualize your personal energy field as a luminous egg-shaped bubble extending about 2-3 feet around your body

See this boundary becoming clear and strong

If you notice any tears, leaks, or weak spots in this field, direct healing light to repair them

State: "My energy field is now strong, clear, and permeable only to what serves my highest good."

Feel the integrity and resilience of your restored energy boundary

Step 6: Frequency Infusion (1-2 minutes)

Complete the practice by filling your cleansed field with your chosen manifestation frequency:

Identify the primary frequency you wish to embody for manifestation (abundance, love, creativity, etc.)

Visualize this frequency as a specific color of light filling your entire field

Feel the qualities of this frequency in your body and emotions

State: "I now vibrate in perfect resonance with [your chosen frequency]."

Allow this frequency to stabilize and settle into your being

Step 7: Gratitude Seal (30 seconds)

Seal the practice with gratitude:

Place your hands over your heart

Express genuine gratitude for this cleansing process

Thank your body and energy field for supporting you throughout the day

Acknowledge yourself for taking time for this important self-care

Variations for Different Needs:

For Emotional Overload:

If you've experienced intense emotions, add this between steps 3 and 4:

Place one hand on your heart and one on your belly

Breathe deeply into these areas

With each exhale, release the emotional energy into the earth

Continue until you feel a noticeable lightening

For Mental Chatter:

If your mind is particularly active, add this between steps 2 and 3:

Gently tap your forehead with your fingertips for 30 seconds

Visualize mental static being discharged through this tapping

Take three deep breaths focused on the space between your

eyebrows

For Physical Exhaustion:

If your body feels depleted, add this after step 6:

Lie down and place your hands palms-up at your sides

Visualize drawing earth energy up through your feet

Simultaneously visualize cosmic energy entering through your crown

See these energies meeting at your heart center, revitalizing your system

For Electromagnetic Clearing:

If you've spent significant time with electronics, add this after step 3:

Rub your hands together vigorously until they feel warm

Shake them out forcefully for 10-15 seconds

Sweep your hands from head to toe, 2-3 inches away from your body

Visualize electromagnetic residue being combed from your field

Quick Version for Busy Evenings (3 minutes):

When time is limited, use this abbreviated version:

Stand in a shower or imagine standing in one

Set your intention for cleansing

As you feel the water (real or imagined) flowing over you, visualize it washing away all energetic debris

State: "Water, cleanse my energy. All that isn't mine or doesn't serve me is now released."

Finish by visualizing yourself surrounded by protective light

Express gratitude for the cleansing

Integration with Physical Showering:

For maximum efficiency, you can integrate this practice with your actual evening shower:

Begin with your intention before stepping into the shower

As you wash your physical body, simultaneously visualize cleansing your energy body

Use the sound of the water to help focus your mind

Assign specific cleansing intentions to different parts of your shower routine:

Shampoo: Cleansing thought patterns

Body wash: Cleansing emotional residue

Rinse: Final release of all energetic debris

Step out of the shower imagining both your physical and energetic bodies completely refreshed

The Energy Shower technique creates a clean vibrational slate for your sleep state, when much of your subconscious manifestation work occurs. By making this practice a non-negotiable part of your evening routine, you ensure that the frequency you carry into sleep supports rather than hinders your manifestation goals.

Just as you wouldn't plant seeds in soil filled with debris, you shouldn't program your subconscious in an energy field cluttered with the day's accumulations.

Gratitude Review

As the day comes to a close, the vibrational state you carry into sleep profoundly affects both your rest quality and your subconscious manifestation work. Of all the frequencies you could cultivate before sleep, gratitude stands as one of the most powerful—creating a state of heart coherence that optimizes your entire system for manifestation.

The Evening Gratitude Review isn't just another item on your spiritual to-do list. It's a scientifically validated practice that creates measurable changes in your heart rhythm patterns, nervous system function, and brain chemistry—all of which directly impact your vibrational broadcast and manifestation abilities.

The Science of Evening Gratitude

Research from the HeartMath Institute has shown that gratitude creates one of the most coherent heart rhythm patterns of any emotional state. This coherence:

Synchronizes communication between your heart and brain

Balances your autonomic nervous system

Reduces stress hormones like cortisol

Increases beneficial hormones like DHEA

Creates an electromagnetic field that extends beyond your body

When you enter sleep in this coherent state, you create optimal conditions for the subconscious processing that occurs during sleep cycles. Your brain literally consolidates learning, processes emotions, and integrates new patterns more effectively when you fall asleep in a state of gratitude.

The Complete Evening Gratitude Review (5-10 minutes)

This structured practice goes beyond simple gratitude lists to create a comprehensive vibrational shift that prepares your entire system for manifestation-aligned sleep.

Step 1: Physical Preparation (1 minute)

Create the right conditions for this practice:

Find a comfortable position, either sitting up in bed or in a chair

Place one hand over your heart center

Take three deep breaths, exhaling completely each time

Gently close your eyes

Feel your body relaxing into the support beneath you

Step 2: Day Reflection (2-3 minutes)

Review your day with a specific focus on moments of blessing:

Mentally scan through your day from morning to evening

Notice moments when:

You experienced joy, peace, or love

You overcame a challenge

You connected meaningfully with others

You made progress toward your goals

You received unexpected support or resources

For each moment, briefly relive the positive feeling associated with it

Acknowledge both significant events and small, easily overlooked blessings

Step 3: Specific Appreciation (2-3 minutes)

Now focus on specific categories of gratitude:

People: Identify 1-3 people who positively impacted your day, directly or indirectly. Feel appreciation for their presence in your life.

Opportunities: Recognize 1-3 opportunities that came your way today, even if they appeared as challenges. Appreciate the growth potential in each.

Personal Qualities: Acknowledge 1-3 of your own strengths or positive qualities that you expressed today. Practice self-appreciation without ego.

Physical Blessings: Notice 1-3 aspects of your physical experience to be grateful for—health, comfort, nourishment, or sensory pleasures.

Manifestation Progress: Identify any signs, however subtle, of movement toward your manifestation goals. Express gratitude for this progress.

For each item, don't just name it—feel the genuine emotion of appreciation in your body, particularly in your heart area.

Step 4: Heart Coherence Amplification (1-2 minutes)

Deepen the coherence created by your gratitude:

Maintain your hand over your heart

Breathe slowly and deeply, imagining your breath flowing in and out through your heart area

With each inhale, draw in the feeling of gratitude more deeply

With each exhale, allow this feeling to expand throughout your body

Visualize your heart emitting a warm, glowing light that grows stronger with each breath

Feel this coherent energy extending beyond your body, creating a field of gratitude around you

Step 5: Manifestation Bridge (1 minute)

Create a bridge between your gratitude state and your manifestation goals:

From this coherent state, bring to mind your primary manifestation goal

Instead of focusing on what you don't yet have, feel genuine gratitude for this manifestation as if it's already complete

State silently or aloud: "I am deeply grateful that [your manifestation] is now coming to me in perfect timing."

Feel the emotional resonance of having already received what you're manifesting

Notice how this perspective of "grateful anticipation" differs from "wanting" or "needing"

Step 6: Sleep Transition (30 seconds)

Prepare to carry this coherent state into sleep:

If you're not already in bed, move there now

As you settle into your sleep position, maintain awareness of the gratitude feeling in your heart

Set the intention: "I sleep in gratitude, and this coherent state continues to attract my manifestations throughout the night."

Release any effort to maintain the practice and allow yourself to drift naturally toward sleep

Variations for Different Personality Types:

For Analytical Types:

If you tend to be more mentally oriented:

Keep a physical gratitude journal by your bed

Write 3-5 specific items with details about why you're grateful for each

Rate your gratitude feeling on a scale of 1-10 for each item

Look for patterns in what consistently generates the strongest gratitude response

For Visual/Creative Types:

If you're more visually oriented:

Create a mental gratitude movie, seeing each blessing as a scene

Visualize each thing you're grateful for surrounded by colored light

Imagine taking photographs of your day's blessings and creating a mental album

Draw or sketch a simple symbol of your day's primary blessing

For Kinesthetic/Feeling Types:

If you're more physically/emotionally oriented:

Place objects that represent the day's blessings on your nightstand

Use physical gestures to express gratitude (hands over heart, arms open wide)

Focus intensely on the bodily sensations associated with gratitude

Express gratitude through gentle movement or stretching before settling into stillness

For Auditory Types:

If you're more sound-oriented:

Speak your gratitude aloud, paying attention to the tone and rhythm of your voice

Record your gratitude list and play it back as you prepare for sleep

Express gratitude through humming or singing a simple melody

Listen to gratitude-themed guided meditations

Overcoming Common Challenges:

For Difficult Days:

When finding gratitude feels challenging:

Begin with the most basic blessings—breath, heartbeat, shelter

Acknowledge the contrast that difficult experiences provide

Express gratitude for your own resilience

Find one small positive aspect within a larger challenging situation

For Sleepiness:

If you're too tired to complete the full practice:

Simplify to just three deep breaths with your hand on your heart

Name one thing you're grateful for with each exhale

Allow the feeling of gratitude to be your last conscious awareness

For Busy Minds:

If your mind is racing with thoughts:

Externalize your gratitude by speaking it aloud

Be very specific and detailed about each gratitude item

Focus on sensory aspects of what you're grateful for

Use the phrase "I notice I'm thinking about..." when your mind wanders, then gently return to gratitude

For Emotional Heaviness:

If you're carrying difficult emotions:

Acknowledge the emotions without judgment

Express gratitude for your capacity to feel deeply

Find gratitude for the smallest positive moments amid the difficulty

Be grateful for the temporary nature of all emotional states

The Evening Gratitude Review creates a vibrational container for sleep that supports your manifestation work on multiple levels. By making this practice the last activity of your day, you ensure that the frequency of gratitude—one of the highest vibrational states accessible to humans—becomes the dominant broadcast of your energy field during the critical hours of sleep.

This practice doesn't just feel good—it creates measurable coherence in your entire system that directly enhances your manifestation capabilities.

Sleep Entrainment

Sleep isn't just rest—it's a powerful time for subconscious manifestation work. During sleep, your conscious mind's resistance is lowered, giving you direct access to the subconscious patterns that govern much of your manifestation ability. By deliberately entraining your brainwaves before and during sleep, you can harness this natural state for profound manifestation programming.

This section explores how to use Theta and Delta brainwave entrainment to optimize your sleep for manifestation and program your subconscious mind while you rest.

The Manifestation Power of Sleep States

Your brain cycles through different brainwave states during sleep, each offering unique opportunities for manifestation work:

Theta State (4-7 Hz):

Occurs during light sleep and dreaming

Creates access to the subconscious mind

Enables vivid visualization and emotional programming

Facilitates creative problem-solving and intuitive insights

Allows for belief reprogramming with reduced resistance

Delta State (0.5-4 Hz):

Dominates during deep, dreamless sleep

Activates the body's healing mechanisms

Enables deep subconscious repatterning

Facilitates integration of new beliefs at the identity level

Creates profound rest that supports high-frequency waking states

By deliberately working with these states, you can transform sleep from passive rest into active manifestation programming.

The Complete Sleep Entrainment Protocol

This comprehensive approach includes pre-sleep preparation, falling asleep with entrainment, and optional all-night programming.

Part 1: Pre-Sleep Theta Programming (15-20 minutes)

This practice uses the hypnagogic state—the transition between wakefulness and sleep—for powerful subconscious programming.

Environment Preparation (2-3 minutes)

Dim lights at least 30 minutes before bedtime

Eliminate blue light exposure from screens

Set room temperature slightly cool (65-68°F/18-20°C is optimal)

Ensure your bedroom is free of clutter and distractions

Optional: Use essential oils like lavender or cedarwood to signal sleep

Physical Relaxation (3-5 minutes)

Lie comfortably in bed, preferably on your back

Take five deep breaths, extending your exhales

Perform a quick body scan, releasing tension from each area

Allow your body to feel heavy and supported by the bed

Let your eyes close naturally when ready

Theta Entrainment Initiation (5-7 minutes)

Begin listening to a Theta entrainment audio (available via QR code or at www.vibrationbook.com/theta)

Use comfortable headphones for optimal effect

Allow the sound to guide your brainwaves into the Theta range

Notice your consciousness beginning to drift while maintaining awareness

If your mind wanders, gently bring it back to the sound

Manifestation Programming (5-7 minutes)

From this receptive Theta state, begin your manifestation programming

Visualize your desired manifestation as already complete

Experience it through all senses—see, hear, feel, taste, and smell it

Generate the emotions associated with this manifestation

Repeat a simple affirmation that encapsulates your manifestation

Allow yourself to drift deeper while maintaining the manifestation image

Sleep Transition

As you feel yourself approaching sleep, release any effort to control the process

Allow your manifestation visualization to be the last thing in your awareness

Drift naturally into sleep while the entrainment audio continues

Part 2: All-Night Delta Programming (Optional)

For deeper subconscious reprogramming, you can use Delta entrainment throughout your sleep cycle.

Setup Requirements

Comfortable sleep headphones designed to be worn all night

Delta entrainment audio with a gradual introduction (available via QR code)

Audio player that can run continuously or on a timer

Optional: Pillow speakers as an alternative to headphones

Programming Preparation

Before starting the Delta audio, clearly state your manifestation intention

Write your intention on a small card and place it under your pillow

Set the audio volume low enough to not disturb sleep but still be audible

Timing Options

Full Night: Set the Delta entrainment to play continuously

First 3 Hours: Set a timer to play during the first deep sleep cycle

Last 2 Hours: Set a timer to begin the audio 2 hours before waking

Morning Integration

Upon waking, remain still for 1-2 minutes

Notice any dreams, insights, or feelings that arose during sleep

Reaffirm your manifestation intention before rising

Record any significant dreams or impressions in your journal

Specialized Entrainment Programs

Different manifestation goals benefit from specific types of sleep entrainment:

For Abundance Manifestation:

Use entrainment embedded with abundance affirmations

Layer in 528 Hz (the "prosperity frequency")

Program visualization of receiving and circulating wealth

Focus on gratitude for abundance already present

For Health Manifestation:

Use entrainment with 432 Hz (the "healing frequency")

Layer in body-specific healing suggestions

Program visualization of cells regenerating during sleep

Focus on feeling vibrant health in your body

For Relationship Manifestation:

Use entrainment with heart-opening frequencies

Layer in self-love affirmations

Program visualization of giving and receiving love

Focus on feeling the emotions of fulfilling connection

For Creative Manifestation:

Use entrainment that alternates between Theta and Alpha

states

Layer in creative problem-solving suggestions

Program questions for your subconscious to solve during sleep

Focus on receiving insights and inspiration

Advanced Techniques for Experienced Practitioners

Once you're comfortable with basic sleep entrainment, explore these advanced approaches:

Lucid Dream Manifestation

Combine sleep entrainment with lucid dreaming techniques to consciously create within the dream state:

Add a lucidity trigger to your pre-sleep affirmations

Program the intention to recognize you're dreaming

When lucid, actively create and experience your manifestation

Upon waking, immediately reinforce the experience as real

Sleep Cycle Targeting

Time different entrainment frequencies to match your natural sleep cycles:

First 90 minutes: Theta entrainment for initial programming

Middle cycles: Delta entrainment for deep integration

Final cycle: Theta/Alpha mix for pre-waking reinforcement

Binaural-Isochronic Combinations

Use more sophisticated audio technology:

Binaural beats for depth of entrainment

Isochronic tones for strength of entrainment

Specific carrier frequencies for targeted effects

Embedded subliminal affirmations for direct subconscious access

Troubleshooting Common Challenges

If You Have Difficulty Falling Asleep with Entrainment:

Reduce the volume to barely audible levels

Use a timer to automatically stop the audio after 30 minutes

Try different types of entrainment sounds (nature-based, pure tones, etc.)

Begin with shorter sessions until you adapt to the process

If You Experience Unusual Dreams:

This is often a sign of subconscious processing and is beneficial

Record these dreams for insights into subconscious blocks

Adjust your pre-sleep affirmations to be more gentle if needed

Express gratitude for this deeper processing

If You Wake During the Night:

Keep the entrainment playing at low volume

Avoid checking the time or using screens

Reaffirm your manifestation intention

Focus on the feeling of gratitude as you return to sleep

If You Don't Remember Dreams:

Place a journal and pen beside your bed

Set the intention to remember dreams before sleep

Remain still upon waking and scan for dream fragments

Record even small impressions, which often lead to fuller recall

Sleep entrainment transforms your eight hours of rest into a powerful manifestation incubator. By deliberately working with your brain's natural sleep cycles and states, you create direct access to the subconscious patterns that govern your manifestation abilities.

This practice doesn't require additional time in your day—it simply optimizes the time you already spend sleeping, turning passive rest into active creation. With consistent use, sleep entrainment can resolve subconscious blocks, install new empowering beliefs, and align your entire being with the frequency of what you wish to manifest.

Your manifestation work doesn't need to stop when you close your eyes. In fact, with proper entrainment, that's when some of the most powerful work begins.

Gratitude Review and Sleep Entrainment Protocols

Your evening routine is equally important as your morning practice—it's when you integrate the day's manifestation work, clear accumulated energy, and program your subconscious mind for overnight manifestation processing.

The Science of Evening Frequency Reset

Why Evening Frequency Work is Critical:

- **Subconscious Programming**: Transition to sleep optimal for subconscious reprogramming
- **Cellular Regeneration**: Sleep rebuilds body according to energetic templates
- **Dream State Manifestation**: Dreams continue manifestation work in quantum realms
- **Emotional Integration**: Evening processing prevents manifestation blocks accumulating
- **Overnight Frequency Maintenance**: Proper practice maintains high vibration during sleep

The Complete Evening Frequency Reset Protocol (25-30 minutes)

Phase 1: Daily Energy Review and Clearing (8-10 minutes)

The Vibrational Day Scan:

Morning Energy Check-In (2 minutes):

- Remember how you felt when you woke up
- Assess how well morning routine aligned your energy
- Thank yourself for morning manifestation work completed
- Feel appreciation for starting day with intention

Midday Energy Assessment (3 minutes):

- Identify moments when energy felt low or scattered
- Acknowledge times when you felt aligned and powerful
- Appreciate any movement breaks or resets completed
- Send compassion to any difficult moments
- Extract wisdom from challenging experiences

Evening Energy Evaluation (3 minutes):

- Honestly assess current energy level and quality
- Acknowledge all forward movement toward manifestations
- Identify any tension or stress still held in body
- Feel satisfaction of another day of conscious living

Energy Clearing Visualization (2-3 minutes):

- Visualize brilliant white light showering over entire energy field
- See light dissolving all heavy, sticky, or foreign energies
- Send dissolved energies into Earth for transformation

- Feel energy field now clear and radiant
- Take three deep breaths feeling completely energetically clean

Phase 2: Gratitude Amplification Practice (7-8 minutes)

The Three-Level Gratitude Review

Level 1: Immediate Gratitude (3 minutes)

- Identify at least 3 ways primary manifestation moved forward today
- Include both obvious progress and subtle signs/synchronicities
- Feel genuine appreciation for each step forward
- Recall surprises, kindnesses, pleasant unexpected events
- Include small joys: good coffee, kind smile, perfect weather
- Appreciate challenges that helped you grow or learn

Level 2: Life Foundation Gratitude (2-3 minutes)

- Thank your body for carrying you through another day
- Appreciate specific functions: breathing, heartbeat, movement
- Feel appreciation for family, friends, colleagues, strangers
- Include pets, plants, other living beings that brought joy
- Feel grateful for shelter, food, safety, freedom, opportunities

Level 3: Universal Gratitude (2 minutes)

- Thank spiritual forces supporting your manifestation
- Appreciate cosmic intelligence orchestrating synchronicities
- Feel gratitude for being part of conscious, creative universe
- Thank ancestors whose choices made your life possible

- Feel appreciation for future self who has achieved manifestations

Gratitude Integration:

- Feel appreciation flowing through every cell of body
- Let gratitude expand heart center until it fills entire energy field
- Feel this gratitude making you magnetic to more blessings
- Send gratitude out to universe as invitation for continued blessings
- "I sleep in gratitude and wake in gratitude. My appreciation draws infinite blessings."

Phase 3: Sleep Entrainment and Subconscious Programming (10-12 minutes)

Progressive Relaxation with Manifestation Focus (4-5 minutes):

- "My toes and feet relax completely, releasing me into manifestation alignment"
- "My legs relax deeply, supporting my manifestation journey"
- "My belly relaxes, digesting today's manifestation progress"
- "My chest and heart relax, opening to receive all heart's desires"
- "My arms and hands relax, releasing control and trusting divine timing"
- "My shoulders and neck relax, releasing all manifestation burden"
- "My head and face relax completely, opening to divine guidance"

Manifestation State Deepening (2 minutes):

- Count backward from 10 with each exhale
- Notice dreamy, floating sensation indicating theta brainwave access
- Feel subconscious mind opening like fertile soil ready for seeds
- Release all conscious effort, trust subconscious to continue work

Subconscious Manifestation Programming (3-4 minutes):

Future Memory Installation:

- In theta state, create vivid "memory" of primary manifestation complete
- Experience through all senses: seeing, hearing, feeling, smelling, tasting
- Include emotional fulfillment of having manifested this desire
- Anchor with: "This is already real in my subconscious mind"

Belief System Updating:

- Identify strongest limiting belief about manifestation
- Install new belief: "I easily and naturally manifest [desire]"
- Repeat 15-20 times, feeling it integrating into subconscious
- See evidence of new belief already showing up in life

Identity Transformation:

- Connect with identity of person who already achieved manifestation
- Feel their confidence, gratitude, natural expectation of success
- Let this identity settle into subconscious as true self

- "I am already the person who has achieved this manifestation"

Sleep Programming and Overnight Activation (2-3 minutes):

Dream Programming:

- "Tonight I dream of my manifestation in perfect detail"
- "My subconscious mind works on my manifestation all night"
- "I wake with perfect clarity about my next manifestation steps"
- "All sleep learning integrates perfectly into waking consciousness"

Overnight Frequency Maintenance:

- "I maintain high manifestation frequency throughout sleep"
- "My energy field attracts manifestation opportunities all night"
- "Every cell in my body aligns with my manifestation during sleep"
- "I co-create with divine intelligence throughout the night"

Morning Integration Setup:

- "I wake refreshed with clear manifestation guidance"
- "My morning routine builds upon tonight's subconscious work"
- "Each night's work accelerates my manifestation timing"
- "I now release all conscious effort and trust the process completely"

Advanced Evening Practices

Extended Evening Ritual (45-60 minutes for weekends):

- **Journaling Integration** (15 minutes): Write detailed manifestation progress
- **Sound Bath** (10-15 minutes): Play 528 Hz during entire routine
- **Advanced Visualization** (10-15 minutes): Create detailed manifestation movie

Quick Evening Reset (10 minutes for busy evenings):

- **5-Minute Clearing**: Brief scan, white light clearing, top 3 gratitudes
- **3-Minute Programming**: Quick relaxation, state manifestation complete
- **2-Minute Integration**: Thank divine intelligence, release with trust

Troubleshooting

"Mind Too Active": Extend clearing phase, use longer progressive relaxation
"Fall Asleep Early": This is ideal—theta naturally transitions to sleep
"No Change Next Morning": Track subtle improvements, notice synchronicities
"Too Tired": Honor body's needs, use quick version on exhausting days

Sleep Environment Optimization

Bedroom Frequency Setup:

- Keep cool (below 70°F) for better theta access
- Minimize electromagnetic fields: WiFi off, phones in airplane mode

- Blackout curtains to eliminate light pollution
- Air-purifying plants for clean oxygen

Pre-Sleep Support:

- No screens 1-2 hours before sleep
- Light stretching to prepare body
- Herbal tea (chamomile, passionflower) for theta transition
- Essential oils (lavender, sandalwood) for spiritual openness

MASTERY & MANIFESTATION

CHAPTER 11: ALIGNING DESIRE WITH FREQUENCY

Desires as Frequency Codes

Are you ready to experience the transformative power of consistent vibrational practice? This 7-Day Frequency Reset Challenge provides a structured program to establish the habit of high-frequency living and create measurable shifts in your manifestation results.

This isn't just another 7-day challenge—it's a progressive system designed to build upon itself day by day, creating a complete reset of your vibrational baseline. Each day focuses on a specific aspect of frequency management while maintaining continuity through core practices.

Preparation:

Before beginning the challenge, gather these materials:

A dedicated journal for tracking your experience

Headphones for audio practices

The companion audio files (available via QR code or at www.vibrationbook.com/reset)

A quiet space where you can practice uninterrupted

Optional: crystals, essential oils, or other vibrational tools that resonate with you

Challenge Structure:

Each day includes:

A morning routine (15-20 minutes)

A midday reset (3-5 minutes)

An evening integration practice (10 minutes)

A specific focus area with unique practices

Tracking prompts for your journal

Day 1: Baseline & Intention

Morning Routine (20 minutes):

Complete the full Breath, Sound, Intention practice from earlier in this chapter

Add 5 minutes of journaling to establish your baseline:

Rate your current vibrational state (1-10)

Note your primary manifestation goal for this reset

Identify your biggest vibrational challenges

Set a clear intention for what you want to experience during this 7-day reset

Midday Reset (3 minutes):

Step away from all devices

Place your hands over your heart

Take 6 deep breaths, focusing on the sensation in your heart center

Repeat your reset intention three times

Return to your activities with renewed awareness

Evening Integration (10 minutes):

Listen to the "Day 1: Baseline" audio meditation

Journal responses to these prompts:

What did you notice about your vibrational state today?

When did your frequency feel highest? Lowest?

What patterns or triggers affected your state?

Set one specific intention for tomorrow

Day 2: Physical Vessel

Focus: Today emphasizes how your physical body affects and carries your vibrational frequency.

Morning Routine (20 minutes):

Basic Breath, Sound, Intention practice (15 minutes)

Add the Physical Frequency Activator (5 minutes):

20 jumping jacks to circulate energy

1 minute of rhythmic spinal movement (gentle twists side to side)

1 minute of humming while tapping your body from feet to head

3 deep breaths with arms raised overhead

Midday Reset (5 minutes):

Step outside if possible

Do 1 minute of energetic breathing (quick inhales, full exhales)

Stretch your body in all directions for 2 minutes

Place your bare feet on the earth for 2 minutes (or visualize this connection if not possible)

Evening Integration (10 minutes):

Listen to the "Day 2: Physical Vessel" audio meditation

Journal responses to these prompts:

How did physical movement affect your vibrational state today?

What foods or substances raised or lowered your frequency?

How did your energy level fluctuate throughout the day?

What one physical practice will you continue beyond this challenge?

Day 3: Emotional Coherence

Focus: Today emphasizes emotional awareness and coherence as keys to vibrational management.

Morning Routine (20 minutes):

Basic Breath, Sound, Intention practice (15 minutes)

Add the Emotional Coherence Practice (5 minutes):

Place both hands over your heart

Recall a moment of pure joy, love, or peace

Breathe this feeling throughout your body

Set the intention to return to this emotional state throughout the day

Midday Reset (3 minutes):

Use the Heart-Focused Breathing technique:

Place attention on your heart area

Breathe as if through your heart

Recall your morning emotional state

Silently repeat: "I choose coherence now"

Evening Integration (10 minutes):

Listen to the "Day 3: Emotional Coherence" audio meditation

Journal responses to these prompts:

What emotions dominated your experience today?

How did you respond to emotional triggers?

When did you feel most emotionally coherent?

What helps you return to emotional balance?

Day 4: Mental Clarity

Focus: Today emphasizes how thought patterns affect your vibrational frequency.

Morning Routine (20 minutes):

Basic Breath, Sound, Intention practice (15 minutes)

Add the Mental Clarity Practice (5 minutes):

Observe your thoughts without attachment for 2 minutes

Write down any recurring thought patterns

Create three clear thought statements that align with your manifestation

Repeat these statements while visualizing them as bright light in your mind

Midday Reset (3 minutes):

The Thought-Stop Technique:

Close your eyes

Say "STOP" firmly in your mind

Take one deep breath

Replace any limiting thoughts with one of your morning clarity statements

Open your eyes with renewed focus

Evening Integration (10 minutes):

Listen to the "Day 4: Mental Clarity" audio meditation

Journal responses to these prompts:

What thought patterns dominated your day?

How did your thoughts affect your vibrational state?

What clarity statements were most effective?

What mental shifts created the biggest vibrational change?

Day 5: Environmental Resonance

Focus: Today emphasizes how your physical environment affects your frequency.

Morning Routine (20 minutes):

Basic Breath, Sound, Intention practice (15 minutes)

Add the Space Clearing Mini-Ritual (5 minutes):

Open a window to allow fresh air circulation

Clap in the corners of your primary living/working space

Light a candle with the intention of elevating the space's frequency

State aloud: "This space now resonates with my highest intentions"

Midday Reset (5 minutes):

Perform a quick environmental adjustment:

Clear any clutter that has accumulated since morning

Add one beauty element (flower, image, crystal)

Play 3 minutes of high-frequency music

Express gratitude for your space

Evening Integration (10 minutes):

Listen to the "Day 5: Environmental Resonance" audio meditation

Journal responses to these prompts:

How did your environment affect your frequency today?

What spaces felt most aligned with your manifestation?

What one environmental change had the biggest impact?

How can you maintain higher environmental frequency?

Day 6: Relational Fields

Focus: Today emphasizes how relationships affect your vibrational state.

Morning Routine (20 minutes):

Basic Breath, Sound, Intention practice (15 minutes)

Add the Relational Field Practice (5 minutes):

Visualize a golden light surrounding you

See this light extending to include those you'll interact with today

Set the intention to maintain your frequency regardless of others' states

Create a mental "frequency shield" for challenging interactions

Midday Reset (3 minutes):

Perform a quick relational reset:

Step away from others briefly

Take 3 deep breaths

Visualize your energy field as distinct but connected to others

Return to interactions with renewed awareness

Evening Integration (10 minutes):

Listen to the "Day 6: Relational Fields" audio meditation

Journal responses to these prompts:

How did others affect your frequency today?

When did you maintain your state despite external influences?

What relationships amplified your frequency?

What boundaries would support your vibrational maintenance?

Day 7: Integration & Commitment

Focus: Today emphasizes bringing all elements together into a sustainable practice.

Morning Routine (20 minutes):

Basic Breath, Sound, Intention practice (15 minutes)

Add the Commitment Practice (5 minutes):

Review your journal entries from the past 6 days

Identify the 3 practices that had the biggest impact

Create a realistic plan to incorporate these into your daily life

Make a spoken commitment to your ongoing frequency management

Midday Reset (5 minutes):

Perform a full-spectrum reset:

1 minute physical movement

1 minute emotional coherence

1 minute mental clarity

1 minute environmental awareness

1 minute relational boundaries

Evening Integration (15 minutes):

Listen to the "Day 7: Integration" audio meditation

Complete the comprehensive review in your journal:

Compare your vibrational baseline from Day 1 to now

Note specific manifestation progress or synchronicities

Identify your biggest insights about your personal frequency

Create your personalized "Frequency Maintenance Plan" with specific daily, weekly, and monthly practices

Write a letter to your future self about maintaining this new vibrational baseline

Tracking Your Results:

Throughout the 7-day challenge, use these tracking methods to document your progress:

Vibrational Scale (Morning and Evening)

Rate your vibrational state on a scale of 1-10 twice daily, noting what affects your ratings.

Manifestation Markers

Record any signs of movement toward your manifestation goals:

Synchronicities related to your desires

Unexpected opportunities

Shifts in your relationship to your goal

Physical evidence of progress

Body Awareness

Note physical sensations that indicate your vibrational state:

Energy levels throughout the day

Areas of tension or openness in your body

Quality of sleep

Digestive function

Overall sense of vitality

Emotional Tracking

Document your emotional patterns:

Predominant emotions each day

Emotional triggers and responses

Recovery time from lower states

Peak emotional experiences

After the Challenge:

To maintain and build upon your 7-day reset:

Create a Sustainable Daily Practice

Identify the elements that had the biggest impact and create a realistic daily practice (even if just 5-10 minutes).

Weekly Frequency Check-In

Schedule a weekly review of your vibrational state and manifestation progress.

Monthly Deep Reset

Choose one day each month for a more comprehensive frequency reset.

Ongoing Education

Continue learning about frequency management through books, courses, or communities.

Environmental Support

Create reminders in your physical space that support your vibrational awareness.

This 7-day challenge provides a structured introduction to conscious frequency management. By the end, you'll have established new neural pathways and energetic patterns that support a higher vibrational baseline—the foundation for more effective manifestation.

Remember that vibrational mastery isn't about perfection but consistency. Even small daily practices, when maintained over time, create significant shifts in your manifestation abilities.

Every desire you hold isn't just a mental concept or emotional longing—it's a specific vibrational pattern, a frequency code that exists in the quantum field. Understanding this fundamental truth transforms how you approach manifestation, moving you from trying to "get" something to aligning with the frequency that naturally attracts it.

This perspective shift is the difference between struggling against the current and flowing with it—between manifestation that feels like hard work and manifestation that unfolds with elegant synchronicity.

The Vibrational Nature of Desire

At its core, every desire represents a specific frequency pattern in the universe. This isn't metaphorical—it's the literal mechanism through which manifestation operates:

Quantum Resonance: Quantum physics has established that everything in existence, at its most fundamental level, is vibrating energy. Your desires exist as specific vibrational signatures within this quantum field.

Emotional Encoding: The feeling associated with your desire —not just the mental concept—creates a specific electromagnetic pattern that your heart and brain broadcast into the field.

Attractor Patterns: These vibrational signatures function as "attractor patterns" that naturally draw similar frequencies toward you through the principle of resonance.

Holographic Reflection: The external manifestation of your

desire is a holographic reflection of its internal frequency pattern—the outer must match the inner according to universal law.

When you truly understand that desires are frequency codes, you stop seeing manifestation as trying to acquire something outside yourself and start recognizing it as a process of vibrational alignment.

The Anatomy of a Desire Frequency

Each desire's frequency code contains multiple components that must align for successful manifestation:

The Essence Frequency

At the heart of every desire is what you're really seeking—the essential feeling or state of being. For example:

The desire for wealth might actually be a desire for freedom, security, or expression

The desire for a relationship might be a desire for connection, validation, or shared joy

The desire for achievement might be a desire for meaning, recognition, or growth

This essence frequency is what you're truly trying to manifest, regardless of the specific form you've attached to it.

The Form Frequency

This is the specific manifestation vehicle you've identified for the essence—the particular job, relationship, amount of money, or experience. The form frequency must be compatible with the essence frequency for manifestation to occur smoothly.

The Worthiness Frequency

This component relates to whether you vibrationally feel aligned with receiving this desire. It's influenced by:

Self-concept and identity

Beliefs about what's possible for you

Past experiences with similar manifestations

Family and cultural programming about deservingness

The Timing Frequency

This relates to the vibrational pattern of when and how the manifestation unfolds:

The natural gestation period for this particular manifestation

Your readiness to receive it

The alignment of other factors necessary for its appearance

The optimal sequence for your overall growth and evolution

The Purpose Frequency

This component connects your desire to its larger meaning and contribution:

How this manifestation serves your soul's journey

How it contributes to others and the greater good

How it fits into your broader life purpose

The growth and evolution it facilitates

When all these frequency components are in alignment, manifestation occurs with remarkable ease and grace. When they're misaligned, you experience the frustration of wanting something that doesn't seem to materialize despite your best efforts.

Reading Your Desire's Frequency Code

Learning to "read" the frequency code of your desires gives you unprecedented manifestation power. Here's how to decode what your desires are really about:

Essence Extraction

To identify the essence frequency of your desire:

Ask yourself: "If I had this, what would it give me that I don't currently have?"

Then ask: "And if I had that, what would it give me?"

Continue this inquiry until you reach a feeling state that doesn't lead to anything else

This final state is the essence frequency of your desire

For example, if you desire a promotion, you might discover through this process that what you really seek is recognition, and beneath that, a sense of worthiness, and beneath that, self-acceptance. Self-acceptance is the essence frequency.

Form Evaluation

To assess whether your chosen form aligns with the essence:

Once you've identified the essence frequency, ask: "Is this specific manifestation the most aligned vehicle for experiencing this essence?"

Consider whether there are other forms that might deliver the essence more effectively

Notice any resistance or incongruence between the form and essence

Sometimes you'll discover that the form you've been pursuing isn't actually the most direct path to the essence you seek.

Worthiness Assessment

To identify worthiness frequency misalignments:

Notice your immediate emotional response when you imagine receiving your desire

Listen for internal voices of doubt, criticism, or warning

Pay attention to physical sensations of contraction or expansion

Observe whether you can fully receive the desire in your imagination

Any discomfort, hesitation, or contraction indicates a worthiness frequency misalignment that needs addressing.

Timing Attunement

To align with the natural timing frequency of your desire:

Ask yourself: "Am I willing to receive this in divine timing, even if it's different from my preferred timeline?"

Notice any attachment to when or how the manifestation must occur

Feel into whether you're genuinely ready for this manifestation now

Consider what preparations or growth might naturally precede this manifestation

Resistance to divine timing creates friction in the manifestation process.

Purpose Connection

To align with the purpose frequency:

Ask: "How does this desire serve my highest evolution?"

Consider: "How might this manifestation benefit others?"

Reflect: "Does this desire feel connected to my deeper purpose?"

Notice whether the desire feels self-contained or part of something larger

Desires connected to purpose carry additional manifestation power through their alignment with universal flow.

Practical Application: Frequency Alignment

Once you understand your desire as a frequency code, you can work directly with that frequency rather than just focusing on actions to "get" what you want:

Essence Embodiment

Instead of waiting to experience the essence until the form manifests, begin embodying the essence frequency immediately:

If the essence is freedom, find ways to experience freedom now

If the essence is connection, cultivate connection in available relationships

If the essence is security, create experiences of security in your current circumstances

This practice closes the perceived gap between you and your desire, accelerating manifestation through resonance.

Form Flexibility

While maintaining clarity about your desired form, practice openness to how it manifests:

Affirm: "This or something better now manifests for my highest good"

Remain curious about unexpected forms that might deliver your essence

Release attachment to specific details that aren't essential to the desire's core

This flexibility allows the universe to deliver your essence through optimal pathways that might be beyond your current imagination.

Worthiness Integration

Directly address any worthiness frequency misalignments:

Identify specific beliefs that create worthiness gaps

Use techniques from earlier chapters to shift these beliefs

Practice receiving in small ways to build your "receiving muscles"

Create evidence of worthiness through intentional experiences

As your worthiness frequency rises, manifestation resistance dramatically decreases.

Divine Timing Surrender

Align with the natural timing frequency of your desire:

Release arbitrary deadlines not based on genuine needs

Trust the perfect unfolding of your manifestation

Focus on the growth and preparation that precedes manifestation

Find joy and meaning in the journey toward manifestation

This surrender paradoxically tends to accelerate manifestation by removing resistance.

Purpose Amplification

Connect your desire to its larger purpose to amplify its manifestation power:

Consider how this manifestation might serve others

Identify how it contributes to your life's mission

Visualize the positive ripple effects beyond your personal benefit

Dedicate some portion of the manifestation to service

This connection to purpose aligns your desire with universal creative forces.

Understanding desires as frequency codes transforms manifestation from a process of acquisition to one of alignment. You stop trying to pull things toward you through force of will and instead adjust your own frequency to naturally attract what resonates with your authentic being.

This perspective shift is the foundation of masterful manifestation—working with the vibrational nature of reality rather than against it.

Why Misaligned Vibes Cancel Manifestations

Have you ever wanted something desperately, done everything "right" to manifest it, yet still found yourself empty-handed? The frustration of this experience is all too common—and it points to one of the most crucial yet overlooked aspects of manifestation: vibrational alignment.

When your conscious desires conflict with your unconscious vibrational patterns, you create cancellation patterns that neutralize your manifestation efforts. Understanding these misalignments and how to resolve them is often the missing piece that transforms struggling manifestation into effortless attraction.

The Physics of Vibrational Cancellation

To understand why misaligned vibrations cancel manifestations, we need to look at the physics principle of wave interference:

When two waves meet, they interact in one of three ways:

Constructive interference: When aligned waves meet, they amplify each other, creating a stronger combined wave

Destructive interference: When misaligned waves meet, they cancel each other out, creating weakness or nullification

Complex interference: When partially aligned waves meet, they create distorted patterns that only partially express either wave

This same principle applies to the vibrational patterns involved in manifestation. When your conscious desires (one wave) meet your unconscious patterns (another wave), they create interference patterns that determine your manifestation results.

The 7 Common Misalignments That Cancel Manifestation

Let's examine the most frequent vibrational misalignments and how they create manifestation interference:

Surface Desire vs. Core Programming

This fundamental misalignment occurs when your conscious desire conflicts with deeper subconscious programming:

You consciously desire financial abundance while subconsciously believing money is scarce or corrupting

You consciously desire a loving relationship while subconsciously believing you're unworthy of love

You consciously desire health while subconsciously identifying with illness

The result is a cancellation pattern where your conscious efforts are continuously undermined by the stronger vibrational broadcast of your subconscious beliefs.

Signs of this misalignment:

Feeling like you're "pushing a boulder uphill" in your manifestation efforts

Experiencing initial progress followed by self-sabotage

Noticing persistent negative thoughts that contradict your stated desires

Feeling internal resistance when visualizing your desire as fulfilled

Stated Goals vs. Energetic Broadcast

This misalignment occurs when your verbal or mental declarations don't match the energy you're actually broadcasting:

You affirm abundance while feeling anxiety about bills

You declare "I am confident" while your body language communicates insecurity

You set goals for health while emotionally resonating with fatigue

Since the universe responds to your energetic broadcast rather than your words, this misalignment creates a confusing interference pattern that yields mixed or negligible results.

Signs of this misalignment:

Others don't seem to take your goals seriously

You feel like you're "faking it" during manifestation practices

Your body feels tense or contracted when stating your affirmations

You attract partial or distorted versions of what you're trying to manifest

Desired Outcome vs. Feared Outcome

This particularly troublesome misalignment occurs when you focus on what you want but simultaneously fear not getting it or fear negative consequences of receiving it:

You focus on attracting a promotion while fearing rejection

You visualize perfect health while worrying about disease

You affirm abundant wealth while fearing the responsibilities it might bring

This creates a direct cancellation pattern where your fear vibration neutralizes your desire vibration.

Signs of this misalignment:

Experiencing anxiety or worry alongside excitement about your desire

Obsessively checking for results or signs of manifestation

Mentally rehearsing explanations for potential failure

Noticing that what you fear often manifests instead of what you desire

Primary Desire vs. Secondary Gains

This subtle misalignment occurs when your conscious desire conflicts with unconscious benefits you receive from the current situation:

You desire weight loss while unconsciously valuing the protection or reduced expectations that come with being overweight

You desire career advancement while unconsciously benefiting from reduced responsibility in your current position

You desire relationship healing while unconsciously gaining freedom or attention from ongoing conflict

The secondary gain creates a contradictory vibrational pattern that interferes with your primary desire's manifestation.

Signs of this misalignment:

Feeling ambivalent or conflicted about your desire

Noticing a pattern of "almost manifesting" but never quite succeeding

Finding yourself creating last-minute obstacles to receiving

Feeling unexpected negative emotions when progress occurs

Individual Desire vs. System Harmony

This misalignment occurs when your personal desire conflicts with the vibrational patterns of larger systems you're part of (family, culture, organization):

You desire financial abundance while your family system has a strong poverty consciousness

You desire authentic self-expression while your cultural system values conformity

You desire work-life balance while your organizational system rewards overwork

The stronger field of the system creates interference with your individual vibrational broadcast.

Signs of this misalignment:

Experiencing resistance or sabotage from those around you

Feeling like an "outsider" when pursuing your desire

Noticing that progress occurs when you're away from certain environments

Experiencing internal conflict between personal desires and group loyalty

Current Desire vs. Past Trauma

This deep misalignment occurs when your present desire triggers unresolved trauma frequencies from your past:

You desire intimate partnership while carrying unhealed abandonment trauma

You desire financial success while holding trauma related to visibility or power

You desire health transformation while trauma is stored in the body

The trauma frequency creates powerful interference that often completely cancels manifestation until healed.

Signs of this misalignment:

Experiencing triggering or activation when visualizing your desire

Noticing strong emotional reactions that seem disproportionate

Finding yourself "freezing" or dissociating during manifestation practices

Experiencing physical symptoms when focusing on certain desires

Conscious Desire vs. Soul Blueprint

This profound misalignment occurs when your ego-driven desire conflicts with your soul's evolutionary plan:

You desire a specific relationship while your soul is calling for a period of independence

You desire wealth through a particular career while your soul is guiding you toward a different path

You desire certain achievements while your soul is focused on inner development

This creates a deep interference pattern where manifestation is blocked not because something is wrong but because something better is aligned with your highest evolution.

Signs of this misalignment:

Encountering persistent obstacles despite doing everything "right"

Receiving intuitive guidance that contradicts your conscious desire

Feeling relief rather than disappointment when manifestation doesn't occur

Noticing synchronicities pointing toward a different direction

Why These Misalignments Are Actually Beneficial

While manifestation cancellation can be frustrating, it actually represents the perfect functioning of universal law. These interference patterns protect you from:

Manifesting contradictions that would create more suffering than satisfaction

Receiving before you're ready to truly hold and integrate what you desire

Creating outcomes that would ultimately lead you away from your highest good

Bypassing necessary healing that serves your long-term evolution

The cancellation isn't a failure of manifestation—it's evidence of a perfectly working system that responds to your total vibrational broadcast, not just your conscious requests.

Resolving Vibrational Misalignments

The key to resolving these cancellation patterns is not trying harder with the same techniques but addressing the specific misalignments:

For Surface Desire vs. Core Programming:

Use techniques from Chapter 4 to identify and shift limiting beliefs

Create new evidence that supports your desired beliefs

Work with theta brainwave states to access and reprogram subconscious patterns

For Stated Goals vs. Energetic Broadcast:

Practice embodiment of the feeling state of your manifestation

Use the techniques from Chapter 2 to achieve heart coherence

Focus on genuine gratitude for elements of your desire already present

For Desired Outcome vs. Feared Outcome:

Acknowledge and process fears rather than suppressing them

Use the Energy Shower technique to clear fear frequencies daily

Practice surrender and trust in the manifestation process

For Primary Desire vs. Secondary Gains:

Consciously identify the benefits of your current situation

Create alternative ways to meet the needs fulfilled by secondary gains

Honor the wisdom of resistance rather than fighting against it

For Individual Desire vs. System Harmony:

Use energy boundaries to maintain your frequency in challenging environments

Consider how your desire might actually serve the larger system

Find or create communities that support your desired frequency

For Current Desire vs. Past Trauma:

Work with qualified support to address and heal trauma patterns

Use gentle, incremental approaches to manifestation that feel safe

Honor the pace that allows integration rather than retraumatization

For Conscious Desire vs. Soul Blueprint:

Practice deep listening through meditation and intuitive practices

Be willing to release or revise desires that create persistent resistance

Trust that what your soul is guiding you toward exceeds your current vision

The path to powerful manifestation isn't forcing your way through resistance—it's creating internal alignment that allows natural, effortless attraction. When you resolve these vibrational misalignments, manifestation transforms from struggle to synchronicity, from effort to ease.

Your manifestation challenges aren't signs of failure—they're precise feedback about where alignment is needed. By addressing these specific misalignments, you remove the interference patterns that have been canceling your manifestations, allowing the clear broadcast of your desires to create corresponding results in your reality.

CHAPTER 12: LIVING A HIGH-VIBE LIFE

Frequency is Contagious

Have you ever noticed how quickly your mood can shift after spending time with certain people? How a room's energy can completely transform when someone specific walks in? This isn't just perception—it's the tangible effect of frequency contagion, one of the most powerful yet underappreciated aspects of vibrational reality.

Your personal frequency isn't just affecting your own manifestation results—it's influencing everyone around you. And conversely, the frequencies of those around you are constantly impacting your vibrational state. Understanding this principle of frequency contagion transforms how you approach both personal manifestation and your relationships with others.

The Science of Energetic Transmission

This isn't just metaphysical theory—it's supported by solid scientific research:

Heart Field Entrainment: The HeartMath Institute has documented how the electromagnetic field generated by one person's heart can directly influence the heart rhythm and brain activity of people nearby. Their research shows that when one person achieves a coherent heart state, the brain waves of others in proximity begin to synchronize with this coherent pattern.

Emotional Contagion: Researchers at the University of California have demonstrated that emotions spread through social networks like a virus. Their studies show that happiness, sadness, and even more complex emotional states can "infect" others up to three degrees of separation away—affecting not just your friends but your friends' friends' friends.

Mirror Neuron Activation: Neuroscientists have identified specialized brain cells called mirror neurons that automatically fire both when we perform an action and when we observe someone else performing the same action. This includes emotional states—when you see someone experiencing joy, your brain partially activates the same neural patterns as if you were experiencing joy yourself.

Limbic Resonance: In their book "A General Theory of Love," researchers Thomas Lewis, Fari Amini, and Richard Lannon describe how our limbic systems (the emotional centers of our brains) automatically attune to those around us, creating a form of emotional synchronization beyond conscious control.

These mechanisms explain why frequency is literally contagious—your vibrational state is constantly being "caught" by others, and you're constantly "catching" the frequencies of those around you.

The Ripple Effect of Your Frequency

Your personal vibrational state creates ripples far beyond your immediate experience:

Direct Entrainment

When you maintain a high frequency state, you directly entrain the energy fields of those in your presence. This happens through several mechanisms:

Their heart rhythms begin to synchronize with yours

Their brainwave patterns shift to match your more coherent state

Their nervous systems calm in response to your regulated state

Their biochemistry changes through mirroring your emotional state

This entrainment happens automatically, without any conscious intention on your part or awareness on theirs. Simply by maintaining your own high frequency, you're helping elevate everyone around you.

Behavioral Modeling

Beyond the direct energetic effects, your frequency creates visible behaviors that others unconsciously adopt:

Your calm response to challenges models emotional regulation

Your solution-focused approach demonstrates possibility thinking

Your grateful perspective highlights what's working rather than what's broken

Your boundary-setting shows healthy self-care and energy management

These behavioral patterns are contagious through social learning mechanisms, creating new possibilities in others' awareness.

Field Effects

Your consistent vibrational state contributes to collective fields that affect larger groups:

Family fields: Your frequency can shift the emotional tone of your entire household

Workplace fields: Your vibration can influence team dynamics and productivity

Community fields: Your energy contributes to neighborhood and community atmospheres

Global fields: Through the butterfly effect, your frequency impacts the collective human field

Research from the Global Consciousness Project at Princeton University has documented how human consciousness creates measurable effects in random event generators, suggesting that our collective frequency literally affects the physical world.

Manifestation Acceleration

When you elevate others' frequencies, you create a powerful manifestation feedback loop:

Their elevated state helps maintain your high frequency

Their manifestation successes reinforce your belief in possibility

Their gratitude and joy contribute to a shared field of abundance

Their growth creates new opportunities that benefit everyone

This creates a virtuous cycle where everyone's manifestation abilities are amplified through mutual frequency elevation.

The Responsibility and Opportunity

Understanding frequency contagion brings both responsibility and opportunity:

The Responsibility:

Recognize that your low-frequency states affect others, not just yourself

Acknowledge that complaining, criticism, and fear are contagious emissions

Take ownership of your energetic impact on family, friends, and colleagues

Commit to frequency management as a form of social responsibility

The Opportunity:

Realize that by elevating your own frequency, you automatically help others

Understand that your personal manifestation work has collective benefits

Recognize your power to shift group dynamics through your individual state

See relationships as opportunities for mutual frequency elevation

Practical Applications of Frequency Contagion

Here's how to work consciously with the contagious nature of frequency:

Strategic Frequency Elevation

Before important interactions, deliberately raise your frequency:

Before family gatherings, elevate your frequency to shift the family field

Before meetings, align with solution frequencies to influence outcomes

Before difficult conversations, establish heart coherence to create safety

Before entering public spaces, strengthen your field to maintain stability

Conscious Entrainment

When someone is in a lower frequency state, use conscious entrainment:

Maintain your higher frequency without trying to "fix" them

Breathe deeply and establish heart coherence in their presence

Hold a compassionate space that allows their frequency to naturally rise

Avoid matching their lower state through unconscious mirroring

Group Field Creation

When working with groups, intentionally establish beneficial frequency fields:

Begin gatherings with practices that synchronize and elevate group frequency

Address discordant energies directly rather than allowing them to affect everyone

Create agreements about the vibrational quality you want to maintain together

Recognize and appreciate frequency elevators within the group

Frequency Protection

In environments where lower frequencies dominate:

Strengthen your personal field through the practices in earlier chapters

Set clear energetic boundaries without judgment or superiority

Use pattern interrupts when you notice yourself being entrained downward

Limit exposure to consistently low-frequency environments when possible

Frequency Leadership

Take conscious leadership of frequency in your spheres of influence:

Be the first to shift from problem-focus to solution-focus

Model frequency reset practices openly, normalizing vibrational management

Create environments that support high-frequency states for everyone

Teach frequency principles to those who are receptive

The Ethics of Frequency Influence

Working with frequency contagion requires ethical awareness:

Respect for Autonomy: While your frequency affects others, respect their sovereign journey. Not everyone is ready to shift frequencies at the same pace or in the same way.

Non-Judgment: Avoid labeling lower frequencies as "bad" or "wrong." All frequencies serve purposes in human experience and development.

Authentic Elevation: Focus on genuine frequency elevation rather than suppressing or denying lower states. Toxic positivity (forcing "good vibes" while denying authentic experience) creates more harm than healing.

Compassionate Presence: Sometimes the most supportive frequency you can offer is compassionate presence with someone in pain, rather than trying to immediately shift them to a "higher" state.

Your Vibrational Legacy

Every interaction leaves an energetic imprint that continues long after the interaction ends. Your frequency creates a vibrational legacy that affects:

The emotional development of children in your life

The belief systems of those who look to you for guidance

The energetic patterns in spaces where you regularly spend time

The collective field of human consciousness

By consciously working with frequency contagion, you create a positive vibrational legacy that continues to influence the world even when you're not physically present.

The contagious nature of frequency means that your personal manifestation practice is never just personal—it's a contribution to collective creation. As you elevate your frequency for your own manifestation goals, you simultaneously help create a world where manifestation becomes easier for everyone.

This perspective transforms manifestation from a potentially self-centered pursuit into a form of service. Your high-vibe life isn't just for you—it's a gift to everyone whose field interacts with yours.

From Practice to Lifestyle

There comes a point in your vibrational journey when conscious practices begin to transform into unconscious patterns —when what once required deliberate effort becomes your natural way of being. This transition from practice to lifestyle represents a quantum leap in your manifestation abilities, as high frequency becomes your default state rather than something you have to constantly work to maintain.

This section explores how to facilitate this crucial transition, addressing the challenges that arise and providing strategies to make high-vibrational living your new normal.

The Four Stages of Vibrational Integration

Understanding the natural progression from conscious practice to embodied lifestyle helps you navigate this transition more effectively:

Stage 1: Conscious Incompetence

In this initial stage, you're aware of vibrational principles but haven't yet developed the skills to consistently apply them:

You notice when your frequency drops but struggle to shift it

You understand concepts intellectually but haven't fully experienced them

You require significant effort and concentration for basic practices

You frequently forget to apply what you've learned in challenging situations

Stage 2: Conscious Competence

As you practice regularly, you develop the ability to manage your frequency, but it still requires deliberate attention:

You can successfully shift your state when you focus on doing so

You remember to apply techniques in most situations

You see consistent results from your practices

You need to consciously monitor your vibrational state

Stage 3: Unconscious Competence

With continued practice, vibrational management begins to become automatic:

You naturally maintain higher frequencies without constant effort

You instinctively apply appropriate techniques when needed

You quickly notice and correct frequency drops

You experience longer periods of sustained high vibration

Stage 4: Embodied Mastery

In this final stage, high frequency becomes your natural state of being:

Your baseline vibration is consistently elevated

You effortlessly maintain coherence even in challenging circumstances

Your presence automatically elevates others' frequencies

Manifestation occurs with minimal resistance as a natural expression of your state

Most people practicing vibrational manifestation hover between Stages 1 and 2, experiencing the benefits of their practices but still requiring constant vigilance. The transition to Stages 3 and 4 represents the shift from practice to lifestyle —from doing high-frequency activities to being a high-frequency person.

The Integration Threshold

Research in habit formation and neuroplasticity suggests that a significant threshold exists in the development of any new pattern. Studies at University College London found that it takes an average of 66 days for a new behavior to become automatic—but this is just the beginning of true integration.

For vibrational living to become your default state, you typically need:

3-6 months of consistent daily practice

Experience applying techniques across diverse life situations

Processing and releasing resistance that emerges during integration

Supportive environmental factors that reinforce the new patterns

This integration threshold explains why many people experience initial benefits from vibrational practices but don't achieve lasting transformation—they haven't sustained their practice long enough to reach the integration point where high frequency becomes self-sustaining.

Challenges in the Transition

Several predictable challenges arise during the transition from practice to lifestyle:

The Regression Pull

As you establish higher vibrational patterns, you'll experience a gravitational pull back to familiar lower states. This regression pull comes from:

Neural pathways that have been reinforced over years or decades

Unconscious comfort with familiar emotional states, even if they're painful

Identity structures built around lower vibrational patterns

Environmental triggers designed to activate lower frequencies

The Authenticity Question

As your frequency rises, you may question whether this new state is "really you" or if you're somehow being fake or inauthentic. This challenge arises from:

Confusing familiarity with authenticity

Attachments to stories about who you are

Fear of losing connection with people who know the "old you"

Uncertainty about navigating relationships from a new frequency

The Integration Backlash

As new vibrational patterns begin to integrate, you may experience a temporary intensification of lower frequencies—similar to how physical toxins can release during a cleanse. This backlash can include:

Unexpected emotional releases

Dreams processing old patterns

Relationship dynamics suddenly highlighting dissonance

Old beliefs surfacing for examination and release

The Plateau Effect

After initial dramatic shifts, you may hit plateaus where progress seems to stall. These plateaus are actually integration periods where:

New neural pathways are being solidified

Subtle aspects of the pattern are being refined

Your system is adjusting to the new energetic baseline

Preparation is occurring for the next level of expansion

Strategies for Successful Integration

These specific strategies facilitate the transition from vibrational practice to lifestyle:

Consistency Through Transition Points

Maintain your practices especially during key transition

points:

The first 3 weeks (initial pattern establishment)

Around day 60-70 (approaching the automaticity threshold)

Between months 3-4 (when regression pull is strongest)

During months 5-6 (when plateau effects often occur)

These critical windows are when most people abandon their practices, just before integration would occur.

Environmental Redesign

Consciously restructure your environment to support your new vibrational baseline:

Create visual reminders of frequency awareness in your daily spaces

Establish "frequency zones" in your home dedicated to specific practices

Reduce or eliminate media that pulls your vibration downward

Reorganize your schedule to prioritize high-frequency activities

Adjust relationships to include more frequency-elevating interactions

Your environment either reinforces or undermines your vibrational state—redesign it to work for rather than against your integration.

Identity Reconstruction

Consciously evolve your identity to incorporate your new vibrational state:

Create new language patterns that reflect your high-frequency identity

Develop personal rituals that mark the transition to your new state

Share your evolving identity with supportive others who can

reflect it back to you

Release attachments to stories about yourself that no longer serve

Consciously choose new self-definitions aligned with your desired frequency

Identity drives behavior at the deepest level—as you shift your core identity, high-frequency living becomes natural rather than effortful.

Nervous System Regulation

Support your nervous system through the integration process:

Ensure adequate rest during intensive vibrational work

Practice grounding activities that help your system integrate higher frequencies

Use physical movement to process energy released during integration

Work with breath practices specifically designed for nervous system regulation

Consider supplemental support like adaptogens, quality nutrition, and hydration

Vibrational shifts create real changes in your nervous system that require physiological support.

Community Anchoring

Leverage the power of group fields to stabilize your new vibrational patterns:

Connect regularly with others practicing similar frequency work

Share your experiences, challenges, and breakthroughs

Participate in group practices that amplify individual efforts

Create accountability partnerships for consistent practice

Celebrate integration milestones together

The field created by aligned groups powerfully supports individual integration.

Cyclical Integration Model

Rather than expecting linear progress, work with natural cycles of integration:

Recognize that expansion is naturally followed by consolidation

Plan for regular periods of intensive practice followed by integration time

Honor the rhythm of challenge and ease in the integration process

Use moon cycles, seasons, or other natural rhythms to structure your practice

Track your cycles to identify your personal integration pattern

This cyclical approach prevents discouragement during necessary integration plateaus.

Manifestation Feedback Loops

Use manifestation results as reinforcement for continued practice:

Document specific manifestations that result from your vibrational work

Notice the correlation between your frequency state and what appears in your reality

Create clear feedback mechanisms that help you recognize cause and effect

Celebrate manifestation successes as confirmation of your integration

Use manifestation challenges as information about areas needing attention

These feedback loops create positive reinforcement that accelerates integration.

The Lifestyle Markers

How do you know when high-frequency living has truly become your lifestyle rather than just your practice? Look for these markers:

Effortless Maintenance: You maintain high frequency without constant attention or effort

Rapid Recovery: When your frequency drops, you return to baseline quickly and naturally

Intuitive Guidance: You receive clear intuitive direction about what supports your frequency

Natural Boundaries: You automatically avoid or limit exposure to lower frequencies

Synchronistic Flow: You experience regular synchronicities and flow states

Accelerated Manifestation: Your manifestations appear with less effort and greater speed

Frequency Leadership: Others naturally look to you for energetic stability and elevation

Identity Integration: Being high-frequency feels like your authentic self rather than a practice

When these markers become your normal experience, you've successfully made the transition from practicing frequency techniques to living a high-frequency life.

This transition represents the ultimate goal of all the practices in this book—not to give you more things to do, but to transform who you are at a fundamental level. When high frequency becomes your natural state, manifestation shifts from something you work at to something that flows through you as the natural expression of your being.

Case Studies & Reader Stories

The principles and practices in this book aren't theoretical—they're proven approaches that have transformed real lives.

These case studies and reader stories illustrate how vibrational work creates tangible results across diverse circumstances and manifestation goals.

Each story highlights different aspects of frequency work, demonstrating that these principles apply universally while manifesting uniquely through each person's journey.

Sarah: From Financial Struggle to Abundance Flow

Sarah, a 42-year-old graphic designer, had struggled with finances her entire adult life. Despite talent and hard work, she constantly lived paycheck to paycheck, with mounting debt that created constant anxiety.

"I tried everything—budgeting apps, financial courses, even visualization techniques for abundance. Nothing seemed to work. I'd make progress, then somehow end up in an even deeper hole."

The turning point came when Sarah realized her vibrational relationship with money was fundamentally misaligned. Through the practices in this book, she discovered several critical frequency mismatches:

Her family had a strong "money is scarce" frequency that she had unconsciously adopted

She carried a belief that financial success would separate her from loved ones

Her body held tension patterns that activated whenever money opportunities appeared

Sarah committed to a comprehensive approach to shift these patterns:

She practiced the Morning Frequency Routine daily, with specific focus on abundance frequencies

She used the Energy Shower technique each evening to clear financial anxiety

She worked extensively with the 528 Hz frequency (the prosperity tone)

She performed the Desire Alignment Ritual specifically focused on her money blocks

The results unfolded in stages:

Months 1-2: Sarah experienced increased awareness of her financial patterns but limited external change. She noticed how often she broadcast scarcity frequencies through her words and emotions.

Months 3-4: Small synchronicities began appearing—unexpected refunds, gifts, and opportunities. Her anxiety about money noticeably decreased, though her circumstances hadn't dramatically changed.

Months 5-6: Sarah received an unexpected offer to design a high-profile project that paid three times her usual rate. She noticed she felt worthy of the opportunity rather than experiencing her usual impostor syndrome.

By Month 8: Sarah had cleared her credit card debt, established savings for the first time in her adult life, and was attracting a steady stream of ideal clients. Most significantly, she reported: "Money isn't a source of stress anymore. It flows in and out with ease, and I trust that what I need will always be there."

The key insight from Sarah's case: Financial manifestation often requires addressing family frequency patterns and body-stored tensions, not just changing thoughts about money.

Marcus: Healing Through Frequency After Chronic Illness

Marcus, a 35-year-old teacher, had battled chronic fatigue syndrome for seven years. Medical interventions provided limited relief, and he had resigned himself to a life of restricted energy and constant health management.

"I was existing, not living. Everything was organized around conserving my limited energy, and I'd given up on ever feeling truly well again."

Marcus discovered vibrational approaches through a friend

and initially was highly skeptical. However, desperate for any improvement, he decided to experiment with frequency work:

He began with sound healing, using specific Solfeggio frequencies daily

He implemented the 3-Breath Reset whenever he noticed energy depletion

He practiced the Shake Technique to release stagnant energy

He used Sleep Entrainment to improve his rest quality

His journey showed a different pattern than Sarah's:

Months 1-2: Marcus experienced immediate but temporary improvements in his energy levels. He noticed that certain practices would give him a few hours of relief before his symptoms returned.

Months 3-4: The improvements began lasting longer, and Marcus identified specific frequency patterns that preceded energy crashes. He developed a personalized "en-ergy forecast system" to predict and prevent downward spirals.

Months 5-8: Marcus experienced his first symptom-free days in years. These gradually increased in frequency, though he still had difficult periods.

By Year 1: While not completely free of his condition, Marcus had regained approximately 70% of normal energy levels—a transformation his doctors considered remarkable. More importantly, his relationship with his body had fundamentally changed.

"I no longer see my body as broken or my enemy. I understand now that energy is a frequency I can influence, not just a limited resource I have to conserve. Some days are still challenging, but I have tools that actually work, and I continue to improve."

The key insight from Marcus's case: Health manifestation often progresses in a "two steps forward, one step back" pattern, and success may look like significant improvement rather than complete transformation, especially with chronic conditions.

Elena & Thomas: Relationship Transformation Through Frequency Work

Elena and Thomas, married for 12 years, had reached a critical point in their relationship. Communication had broken down, intimacy was rare, and both were considering whether separation might be inevitable.

"We loved each other, but we couldn't seem to connect anymore," Elena explained. "Every conversation turned into either an argument or cold distance. We'd tried traditional couples therapy with limited success."

As a last resort, they agreed to try the vibrational approaches in this book together:

They practiced the Evening Gratitude Review as a shared ritual

They used the High-Vibe Circle Audit to identify relationship patterns

They implemented frequency resets before difficult conversations

They created a "frequency agreement" for how to handle triggering situations

Their progress followed yet another pattern:

Weeks 1-3: Initial resistance and skepticism, particularly from Thomas. The practices felt awkward and sometimes increased tension as they highlighted how disconnected they had become.

Weeks 4-8: Gradual softening as they began to recognize how their individual frequencies affected their interactions. They developed a shared language around energy and vibration that allowed them to discuss triggers without blame.

Months 3-4: Breakthrough moments where they could feel the tangible shift in their shared field when they both maintained higher frequencies. Intimacy began returning naturally without force or scheduling.

By Month 6: Their relationship had transformed from a source of stress to a source of support. They reported not just returning to their earlier connection but establishing a deeper bond than they'd ever experienced before.

Thomas reflected: "I was extremely skeptical about all this 'energy' talk at first. But I can't argue with results. We can now feel when our field gets disrupted, and we have tools to reset it together. Problems that used to last for days now resolve in minutes."

The key insight from Elena and Thomas's case: Relationship manifestation requires both partners to take responsibility for their individual frequencies while also learning to manage their shared field as a co-created reality.

Jamal: Career Alignment Through Vibrational Authenticity

Jamal, a 29-year-old marketing executive, was outwardly successful but inwardly miserable. Despite rapid advancement in his corporate career, he felt increasingly disconnected from his work and purpose.

"I was doing everything I thought I was supposed to do—climbing the ladder, making good money, building my reputation. But I felt empty and exhausted all the time. Something fundamental wasn't aligned."

Jamal began exploring frequency work with a focus on career manifestation:

He used the Desire Alignment Ritual to uncover his true calling

He practiced embodying his authentic frequency rather than the corporate persona he'd created

He implemented daily Sound Science techniques to clear decision-making blocks

He used the Vibrational Architecture principles to transform his workspace

Jamal's manifestation journey took an unexpected path:

Months 1-2: Increasing discomfort in his current role as he became more aware of the frequency mismatch. Rather than immediately improving, his situation initially felt worse as his awareness heightened.

Months 3-4: Clarity emerged about his authentic path—combining his marketing expertise with his passion for environmental sustainability, something he'd never considered viable.

Months 5-6: Synchronistic connections appeared, introducing him to key people in the sustainability sector. An unexpected reorganization at his company created an opening to propose a new green initiative.

By Month 9: Jamal had created an entirely new position within his company, heading their sustainability marketing division. This role combined his skills with his passion, at the same compensation level but with dramatically increased fulfillment.

"What amazed me was that I didn't have to leave my company or take a pay cut to find alignment. Once I got clear on my authentic frequency and started broadcasting it consistently, the external reality rearranged itself to match. The position I now have didn't even exist a year ago—it formed around my frequency."

The key insight from Jamal's case: Career manifestation often requires allowing a period of increased discomfort as you align with your authentic frequency, before external reality can reorganize to match your new vibration.

Maya: Manifesting Community Through Frequency Leadership

Maya, a 51-year-old recent empty-nester, found herself isolated and purposeless after her children left home. Having dedicated decades to family life, she had few close connections outside her immediate family and no clear direction.

"I felt invisible and irrelevant. I knew I needed community, but I had no idea how to create it at my age, especially being naturally introverted."

Maya began applying vibrational principles with a focus on manifesting meaningful community:

She practiced the concepts from "Frequency is Contagious" to understand how she could influence group fields

She used the 1-Song Vibration Lift before social situations to manage her anxiety

She applied the principles from "Living a High-Vibe Life" to maintain consistency

She worked with the Solfeggio frequencies related to connection and communication

Her community manifestation unfolded organically:

Months 1-2: Maya focused primarily on her own frequency, with little external action. She worked on clearing blocks around visibility and worthiness.

Months 3-4: She began taking small steps, joining a local gardening club and a book group. Rather than focusing on what she could get from these groups, she consciously attended to the frequency she contributed.

Months 5-8: As she consistently maintained a high-frequency presence, she noticed others naturally gravitating toward her energy. Without deliberately seeking leadership, she found people asking for her input and guidance.

By Year 1: Maya had become the unexpected heart of a thriving community. She hosted regular gatherings in her home, had initiated a neighborhood sustainability project, and had developed deeper friendships than she'd experienced in decades.

"What's fascinating is that I never set out to create or lead anything. I simply focused on maintaining a clear, open frequency and being fully present with whoever was in front of me. Everything else grew organically from that foundation. I now have more meaningful connection than I know what to do with!"

The key insight from Maya's case: Community manifestation often works best through frequency leadership rather than direct seeking or creating. By embodying the frequency you wish to experience in community, you naturally attract and catalyze it around you.

Common Threads Across Manifestation Journeys

While each of these stories reflects unique circumstances and goals, several important patterns emerge across successful vibrational manifestation:

The Consistency Factor: All successful manifestors maintained their practices through the initial period where results weren't yet visible externally.

The Awareness Before Change Pattern: In most cases, awareness of misalignments increased before external circumstances improved, creating a temporary period of greater discomfort.

The Synchronicity Sequence: Small synchronicities typically preceded larger manifestations, serving as feedback that frequency alignment was improving.

The Identity Shift Requirement: Lasting manifestation required not just changed thoughts or actions but fundamental shifts in how the person identified themselves.

The Surrender Paradox: In each case, a point of surrender—releasing attachment to specific outcomes—paradoxically accelerated the manifestation process.

The Frequency First Principle: Attempting to change external circumstances without first addressing frequency created limited or temporary results.

The Integration Timeline: Significant manifestations typically began appearing in months 3-6, after the initial frequency work had time to integrate and stabilize.

These patterns suggest that vibrational manifestation follows predictable stages, even while the specific content of each person's journey remains unique. Understanding these common threads can help you navigate your own manifestation process with greater patience and trust, especially during the initial phases where external evidence may be limited.

The diverse nature of these case studies—spanning financial, health, relationship, career, and community manifestations—demonstrates that frequency principles apply universally across all areas of human experience. The same fundamental laws govern all types of manifestation, though they express through the unique circumstances and purposes of each individual life.

Your own manifestation journey will write its unique story, but it will follow these same universal principles of frequency alignment and expression.

Lifestyle Integration Practices and High-Vibe Maintenance Protocols

Living a high-vibration life means making your manifestation practice seamlessly integrated with your daily existence rather than compartmentalized into specific practice times.

Daily High-Frequency Living Protocols

Morning Integration: Extending High Vibration Beyond Formal Practice

Hydration as Frequency Medicine:

- Drink room temperature water with sea salt and lemon upon waking
- Bless water by holding 30 seconds while feeling gratitude
- See water carrying high-frequency information to every cell
- State: "This water carries my manifestation energy through-

out my body"

Clothing as Frequency Amplifier:

- Choose colors supporting day's primary manifestation intention
- Natural fabrics maintain better energy flow than synthetics
- Wear clothes that feel good—physical comfort supports energetic flow
- While dressing, visualize clothing as energetic armor maintaining high vibration

Technology Boundary Setting:

- Phone-free first hour before morning practice completion
- If checking messages, do with blessing intention rather than reactivity
- Choose information that supports manifestation goals only

High-Frequency Food and Nutrition

Eating as Vibrational Practice:

- **Fresh and Living**: Prioritize fresh fruits, vegetables, minimal processing
- **Gratitude Before Eating**: Bless food and thank all beings involved
- **Mindful Consumption**: Eat slowly, tasting and appreciating each bite
- **Energy Scanning**: Notice how different foods affect you energetically

Manifestation-Supporting Nutrition:

- **Brain Food**: Omega-3 rich foods support clarity and visualization

- **Grounding Foods**: Root vegetables support stability and practical steps
- **Heart-Opening Foods**: Fresh fruits and green vegetables support love-based manifestation
- **Hydration**: Adequate water supports energy flow and cellular processes

Workday High-Frequency Integration

Conscious Communication as Frequency Practice:

1. **Speak from Abundance**: Frame conversations in possibilities rather than problems
2. **Listen with Presence**: Give full attention as manifestation of respect
3. **Express Appreciation**: Regularly acknowledge colleagues' contributions
4. **Solution Orientation**: When problems arise, immediately shift focus to solutions
5. **Energetic Blessing**: Silently send good wishes to difficult people

Work as Service to Manifestation:

- **Purpose Connection**: Find ways work serves larger manifestation goals
- **Skill Development**: Approach challenges as opportunities to develop qualities
- **Financial Gratitude**: Appreciate income as abundance channel
- **Professional Growth**: See career advancement as manifestation practice

Workspace as Manifestation Sanctuary:

- Include small manifestation reminders (crystals, affirmation cards)
- Add living plants to increase life force energy
- Maintain clean, organized space supporting clear thinking
- Use natural light when possible; add full-spectrum bulbs

Evening Lifestyle Integration

Entertainment as Frequency Choice:

- Choose books, movies, podcasts that inspire rather than drain
- Listen to music enhancing emotional state and manifestation feelings
- Limit social media exposure to negative content
- Stay informed without absorbing fear-based news energy

Social Interactions as Manifestation Practice:

- Prioritize time with people supporting growth and manifestation goals
- Steer conversations toward positive, solution-focused, inspiring topics
- Lovingly limit time with people consistently draining energy
- Look for ways to serve others aligning with manifestation goals

High-Frequency Relationship Practices

Family Relationships:

- Daily appreciation for family members, even during challenges
- Address disagreements with intention to understand and

find harmony

- Share appropriate aspects of manifestation journey with supportive members
- Maintain loving boundaries with family not supporting growth

Romantic Relationships:

- Include partner in manifestation discussions when appropriate
- Maintain personal practice while supporting partner's individual path
- Use relationship dynamics as mirror for manifestation blocks and progress
- Practice unconditional love as training for manifesting love-based desires

Friend and Social Relationships:

- Cultivate friendships with others interested in conscious living
- Create reciprocal support with friends encouraging manifestation goals
- Choose social activities enhancing rather than depleting energy
- Notice how different friends affect energy and manifestation confidence

Environmental High-Frequency Living

Home as Manifestation Sanctuary:

- Regular removal of items not serving current manifestation goals

- Surround yourself with beauty that uplifts and inspires
- Include plants, natural materials, nature photography in decor
- Maintain dedicated spaces for manifestation practice and spiritual connection

Daily Environmental Practices:

- Brief appreciation and blessing of living space each morning
- Simple clearing of daily energy accumulation before sleep
- Seasonal adjustments supporting different manifestation cycles
- Welcome guests with intention to share high-frequency energy

Seasonal High-Frequency Living

Spring Integration (March-May):

- Initiate new practices, relationships, projects supporting goals
- Physical, emotional, environmental cleansing for new manifestations
- Nutrition and activities supporting increased energy and growth
- Begin new manifestation cycles aligned with natural growth energy

Summer Integration (June-August):

- Full expression of gifts and talents supporting manifestation goals
- Increased social interaction and community engagement
- Bold action steps toward major manifestation goals

- Prioritize joy and pleasure as manifestation magnets

Autumn Integration (September-November):

- Increased focus on appreciation for manifestations received
- Celebrate and integrate achievements and manifestation successes
- Prepare for winter's introspective energy through lifestyle adjustments
- Extract and integrate wisdom from year's manifestation experiences

Winter Integration (December-February):

- Lifestyle supporting introspection, planning, spiritual development
- Adequate rest and self-care to regenerate manifestation energy
- Use quiet winter energy for developing next year's visions
- Increased spiritual practice during contemplative season

Creating Your Personal High-Frequency Lifestyle Design

Weekly High-Frequency Planning:

Sunday Planning Ritual:

- Review upcoming week for manifestation opportunities and challenges
- Set intention for maintaining high frequency throughout week
- Identify potential energy drains and develop alignment strategies
- Plan specific manifestation actions to integrate with weekly

activities

Daily Check-Ins:

- **Morning**: How can I maintain high frequency throughout today's activities?
- **Midday**: How is my frequency? What adjustment do I need?
- **Evening**: How did I maintain/lose high frequency? What did I learn?

Monthly Lifestyle Assessment:

High-Frequency Lifestyle Review:

- Which activities consistently raise my frequency?
- Which activities consistently drain my energy?
- How can I increase frequency-raising activities?
- How can I eliminate or transform frequency-draining activities?
- What lifestyle changes would most support manifestation goals?

Daily High-Frequency Checklist:

Morning Questions:

- Did I wake naturally feeling rested?
- Am I eating foods that energize rather than drain me?
- Are my first interactions positive and uplifting?
- Do my morning activities align with manifestation goals?

Evening Questions:

- Did I maintain awareness of my energy throughout the day?
- How did I contribute to others' well-being today?
- What manifestation progress did I make through daily activities?

- Am I ending the day feeling fulfilled and aligned?

Weekly Integration Score (Rate 1-10):

- **Nutrition**: Food choices support high frequency
- **Movement**: Physical activity enhances energy
- **Relationships**: Interactions are positive and supportive
- **Work**: Professional life aligns with manifestation goals
- **Environment**: Living space supports high frequency
- **Entertainment**: Media consumption inspires rather than drains
- **Service**: Contributing to others' well-being

Score Interpretation:

- 56-70: Excellent integration of high-frequency lifestyle
- 42-55: Good integration with room for improvement
- 28-41: Moderate integration, focus on 2-3 key areas
- Below 28: Begin with one area for focused improvement

CHAPTER 13: THE SCIENCE OF MIRACLES

Spiritual-Skeptic Bridge

In the space between blind faith and rigid skepticism lies a fertile middle ground where the seemingly miraculous aspects of manifestation can be understood through the lens of emerging science. This bridge between spiritual wisdom and scientific understanding offers a path for both the devoted believer and the analytical skeptic to approach manifestation with integrity and effectiveness.

The "miraculous" results of advanced manifestation work—synchronicities that defy probability, healings that confound medical explanation, manifestations that appear to bend the rules of time and space—need not be relegated to either supernatural mystery or dismissed as coincidence. A growing body of scientific research offers frameworks that help explain these phenomena without diminishing their wonder.

Quantum Mechanics and Non-Local Consciousness

Quantum physics has revealed a reality far stranger and more malleable than classical Newtonian physics suggested. Several quantum principles directly relate to manifestation phenomena:

The Observer Effect

At the quantum level, the act of observation affects what is being observed. Experiments like the famous double-slit experiment demonstrate that particles behave differently when observed versus unobserved. This suggests a fundamental relationship between consciousness and physical reality that traditional science has struggled to explain.

How this relates to manifestation: Your conscious attention literally affects the probability waves of potential outcomes. When you focus consistently on a specific manifestation, you're not just thinking about it—you're interacting with its quantum potential, increasing the probability of that potential collapsing into physical reality.

Non-Locality and Entanglement

Quantum entanglement demonstrates that once connected, particles remain instantaneously connected regardless of distance—what Einstein called "spooky action at a distance." This non-local connection defies classical understanding of space and time.

How this relates to manifestation: The synchronicities and "coincidences" that occur during manifestation work may operate through non-local connections in consciousness. Your intention creates entanglement between your consciousness and the people, opportunities, and circumstances related to your manifestation, allowing connections that appear miraculous from a classical perspective.

The Field Perspective

Quantum field theory suggests that particles are excitations in underlying fields that permeate all of space. Rather than separate objects interacting, reality consists of interconnected fields in constant relationship.

How this relates to manifestation: Your thoughts, emotions, and intentions create excitations in consciousness fields that extend beyond your physical body. These field effects can influence seemingly separate people and events through resonance rather than direct physical causation.

Biofield Science and Electromagnetic Human Emissions

Beyond quantum physics, research into human biofields provides another bridge between spiritual and scientific understandings of manifestation:

Measurable Heart Fields

As discussed earlier, the HeartMath Institute has documented that the human heart generates an electromagnetic field that extends several feet beyond the body and changes pattern based on emotional states. This field can be measured with sensitive magnetometers and has been shown to affect other biological systems.

How this relates to manifestation: Your emotional state creates literal electromagnetic broadcasts that interact with your environment. The heart coherence practices in this book aren't just metaphorical—they're creating measurable changes in your electromagnetic field that affect what you attract and repel.

Biophoton Emissions

Research pioneered by Fritz-Albert Popp has demonstrated that all living cells emit weak light called biophotons. These emissions change based on the health and state of the organism and appear to play a role in cell-to-cell communication.

How this relates to manifestation: Your thoughts and emotional states affect your biophoton emissions, creating a literal "light signature" that varies with your internal state. Advanced manifestation may involve coherent biophoton emissions that create resonance with desired outcomes.

Biofield Therapies Research

Clinical studies on practices like Reiki, Therapeutic Touch, and other biofield therapies have shown measurable effects on healing outcomes, suggesting that human energy fields can directly influence physical processes in both the practitioner and recipient.

How this relates to manifestation: The same principles that allow biofield therapies to affect physical healing may explain how focused intention affects manifestation outcomes. You're not just thinking about what you want—you're directing biofield energy toward specific outcomes.

Neuroplasticity and the Malleable Brain

Neuroscience research on brain plasticity offers another bridge between spiritual and scientific perspectives:

Experience-Dependent Neuroplasticity

Research has conclusively demonstrated that the brain physically changes in response to experience and focused attention. Neural networks strengthen or weaken based on use, creating physical changes in brain structure.

How this relates to manifestation: The visualization and emotional practices in this book literally rewire your neural networks to become more receptive to your desired manifestations. You're not just imagining different outcomes—you're building the neurological architecture to perceive and respond to manifestation opportunities.

The Predictive Brain Model

Current neuroscience suggests that the brain primarily functions as a prediction machine, constantly generating expectations about reality and filtering perception to match these predictions.

How this relates to manifestation: By consistently imagining your manifestation as complete, you're reprogramming your brain's predictive filters to notice opportunities and connections related to your desire. Many manifestation "miracles" involve perceiving what was already available but previously filtered out by your attention systems.

Mirror Neuron Networks

The discovery of mirror neurons—brain cells that fire both when performing an action and when observing that same action—has revolutionized our understanding of empathy, learning, and social connection.

How this relates to manifestation: Through mirror neuron activation, you can literally "rehearse" the experience of having manifested your desire, creating neural patterns identical to those that would form if the manifestation were already physical. This neural rehearsal primes your system to recognize and act on manifestation opportunities.

Epigenetics and Belief-Driven Biology

Research in epigenetics—how genes are expressed or suppressed based on environmental factors—provides another scientific framework for understanding manifestation:

Belief Effects on Gene Expression

Studies have shown that psychological states can trigger epigenetic changes, altering which genes are expressed without changing the underlying DNA sequence.

How this relates to manifestation: Your beliefs about what's possible literally affect your biology at the genetic level. Manifestation practices that shift limiting beliefs aren't just changing your thoughts—they're changing which genes are expressed in your cells.

Transgenerational Epigenetic Inheritance

Research suggests that some epigenetic changes can be passed to subsequent generations, creating inherited tendencies without altering the genetic code.

How this relates to manifestation: The family patterns discussed in earlier chapters may operate partly through epigenetic inheritance. Your manifestation work may be addressing not just your personal limitations but inherited epigenetic patterns from your ancestral line.

Placebo Response Mechanisms

The placebo effect—once dismissed as "just psychological"—is now understood to involve real physiological changes, including altered brain activity, immune function, and hormone production.

How this relates to manifestation: The same mechanisms that make placebos effective may explain how belief and expectation influence manifestation outcomes. Your conviction that manifestation works creates real physiological changes that affect how you perceive and interact with opportunities.

Bridging Perspectives: Neither Blind Faith Nor Closed Skepticism

This scientific framework offers a middle path that honors both spiritual wisdom and scientific inquiry:

For the spiritually inclined, these scientific perspectives don't reduce manifestation to mere mechanics—they provide validation that spiritual practices affect real physical systems in measurable ways.

For the skeptically minded, these frameworks offer evidence-based explanations for manifestation phenomena without requiring belief in supernatural forces outside natural law.

The most powerful approach to manifestation embraces both perspectives:

Use spiritual practices with the understanding that they create measurable changes in your physical and energetic systems

Approach scientific explanations with wonder at the remarkable interconnectedness they reveal

Recognize that our scientific understanding is still evolving, with much yet to be discovered about consciousness and its relationship to physical reality

The "miracles" of manifestation—the synchronicities, the healings, the improbable opportunities—need not be relegated to either blind faith or dismissed as coincidence. They represent the natural operation of subtle but real processes at the intersection of consciousness and physical reality.

This bridge between spiritual and scientific understanding doesn't diminish the magic of manifestation—it enhances it by revealing the elegant mechanisms through which consciousness interacts with the physical world. The universe becomes not less miraculous through this understanding, but more so—a remarkably responsive system designed for conscious creation.

Your Next 90 Days

You've absorbed a wealth of information about vibrational manifestation throughout this book. Now it's time to transform that knowledge into a structured practice that will create tangible results in your life. This 90-day roadmap provides a comprehensive implementation plan that builds progressively, allowing you to develop mastery one step at a time.

The plan is designed to be realistic yet transformative, with each month building upon the previous one. Rather than trying to implement everything at once, this approach allows for proper integration and sustainable progress.

Month 1: Foundation (Days 1-30)

The first month focuses on establishing core practices and baseline awareness.

Week 1: Vibrational AwarenessDaily Practice (15-20 minutes):

Morning: 5-minute breath awareness

Midday: 3-Breath Reset twice daily

Evening: 5-minute gratitude review

Track your vibrational state on a 1-10 scale three times daily

Weekly Focus:

Complete the High-Vibe Circle Audit (Chapter 7)

Identify your three most common frequency drains

Create a simple plan to address one frequency drain

Key Mindset:

Awareness precedes change. This week is about honest observation without judgment.

Week 2: Physical Vessel PreparationDaily Practice (20-25 minutes):

Morning: 10-minute Morning Frequency Routine (Chapter 8)

Midday: One 5-minute Movement Break (Chapter 9)

Evening: 7-minute Energy Shower (Chapter 10)

Drink half your body weight (in pounds) in ounces of water daily

Weekly Focus:

Perform one Space Clearing Ritual (Chapter 6)

Remove one category of frequency-lowering foods from your diet

Create a dedicated space for your vibrational practices

Key Mindset:

Your physical body is the instrument through which frequency expresses. Preparing this instrument is essential foundation work.

Week 3: Emotional CoherenceDaily Practice (25-30 minutes):

Morning: 15-minute Morning Frequency Routine with emphasis on emotional priming

Midday: 1-Song Vibration Lift when needed

Evening: 10-minute Evening Gratitude Review

Practice heart coherence for 2 minutes before challenging situations

Weekly Focus:

Listen to the 528 Hz frequency for 15 minutes daily

Begin tracking emotional patterns and their effects on manifestation

Identify your top three emotional triggers and create reset plans for each

Key Mindset:

Emotional states are frequency broadcasts. Learning to manage these broadcasts is essential for manifestation.

Week 4: Integration and AssessmentDaily Practice (25-30 minutes):

Continue Week 3 practices

Add one 5-minute frequency check-in before bed

Review your manifestation journal daily

Notice and record any synchronicities or shifts

Weekly Focus:

Complete a full Month 1 review in your journal

Identify what practices had the most impact

Set specific intentions for Month 2

Celebrate your consistency and progress

Key Mindset:

Integration is as important as action. This week consolidates your foundation before building upon it.

Month 2: Amplification (Days 31-60)

The second month intensifies your practice and begins active manifestation work.

Week 5: Sound TechnologyDaily Practice (30-35 minutes):

Morning: 15-minute Morning Frequency Routine

Midday: Two 3-Breath Resets and one Movement Break

Evening: 15-minute brainwave entrainment session (Chapter 4)

Continue tracking your frequency throughout the day

Weekly Focus:

Experiment with the 15-Minute Focus Playlist daily

Try one new Solfeggio frequency each day

Create a personalized sound healing playlist

Practice the 108 Oms once this week (weekend recommended)

Key Mindset:

Sound is a direct path to frequency modulation. This week leverages sound technology for deeper shifts.

Week 6: Desire ClarificationDaily Practice (30-35 minutes):

Continue Week 5 practices

Add 5 minutes of specific visualization for your primary manifestation

Begin using manifestation-specific affirmations during your morning routine

Notice resistance that arises when focusing on your desire

Weekly Focus:

Complete the Desire Alignment Ritual (Chapter 11)

Create a vision board or visual representation of your manifestation

Write a detailed description of your manifestation as if already complete

Identify any misalignments between conscious desires and subconscious patterns

Key Mindset:

Clarity creates power. This week focuses on refining exactly what you're calling into your experience.

Week 7: Environmental HarmonizationDaily Practice (30-35 minutes):

Continue Week 6 practices

Add one environmental frequency reset each day

Place manifestation reminders in your primary spaces

Practice maintaining your frequency in challenging environments

Weekly Focus:

Declutter one major area of your home

Apply Feng Shui or Vastu principles to your primary living space

Create a dedicated manifestation altar or focus point

Assess and adjust the frequency of objects in your environment

Key Mindset:

Your environment is either supporting or undermining your frequency. This week creates external conditions that reinforce your manifestation.

Week 8: Integration and AmplificationDaily Practice (30-35 minutes):

Continue Week 7 practices

Increase visualization time to 10 minutes

Add one new frequency tool from any previous chapter

Begin recording evidence of manifestation movement

Weekly Focus:

Complete a full Month 2 review

Assess progress on your specific manifestation

Refine your practice based on what's working best

Set intentions for Month 3

Key Mindset:

Amplification comes through consistent practice and refinement. This week optimizes your approach based on personal results.

Month 3: Mastery (Days 61-90)

The final month focuses on integration, advanced techniques, and establishing vibrational living as your new normal.

Week 9: Subconscious ReprogrammingDaily Practice (35-40 minutes):

Morning: 20-minute Morning Frequency Routine with deeper emotional work

Midday: Customized frequency maintenance based on personal patterns

Evening: 15-minute Sleep Entrainment preparation

Use theta state programming before sleep

Weekly Focus:

Work specifically with limiting beliefs related to your manifestation

Practice the worthiness integration techniques from Chapter 11

Create personalized affirmations based on your specific blocks

Record and work with dreams that relate to your manifestation

Key Mindset:

The subconscious mind is the gatekeeper of manifestation. This week focuses on aligning your deeper mind with your conscious desires.

Week 10: Frequency LeadershipDaily Practice (35-40 minutes):

Continue Week 9 practices

Add conscious frequency management in interactions with others

Practice maintaining your state regardless of external conditions

Notice your impact on group fields

Weekly Focus:

Apply the principles from "Frequency is Contagious" in daily life

Practice frequency leadership in one challenging relationship

Create a strategy for maintaining your state in lower-frequency environments

Begin sharing what you've learned with receptive others

Key Mindset:

Mastery includes the ability to maintain your frequency while engaging fully with the world. This week focuses on real-world application.

Week 11: Manifestation AccelerationDaily Practice (35-40 minutes):

Continue Week 10 practices

Increase focus on evidence of manifestation movement

Practice gratitude for manifestation as if already complete

Release attachment to specific outcomes while maintaining clear intention

Weekly Focus:

Review and refine your manifestation vision based on new insights

Create a "manifestation evidence journal" documenting all signs of movement

Practice the divine timing surrender techniques from Chapter 11

Identify and release any remaining resistance

Key Mindset:

Manifestation accelerates through the paradox of clear intention combined with complete surrender. This week balances these complementary forces.

Week 12: Lifestyle IntegrationDaily Practice (35-40 minutes):

Continue Week 11 practices

Begin designing your ongoing practice beyond the 90 days

Identify your non-negotiable core practices

Create systems for maintaining consistency long-term

Weekly Focus:

Complete a full 90-day review

Document your journey and transformations

Create your personalized ongoing practice plan

Celebrate your commitment and progress

Key Mindset:

The goal is not perfect practice but consistent integration into your lifestyle. This week creates the bridge to ongoing vibrational living.

Implementation Guidelines:

To get the most from this 90-day plan:

Start where you are: If certain practices feel too advanced, modify them to match your current capacity. It's better to do a simplified version consistently than to attempt the full practice inconsistently.

Prioritize consistency over perfection: A 70% practice done daily is more effective than a 100% practice done occasionally. If time is limited, reduce the duration but maintain the frequency of practice.

Track your journey: Keep a dedicated journal for this 90-day process. Document not just what you do but what you experience, notice, and learn.

Create accountability: Share your commitment with someone supportive, join an online community, or work with a coach or mentor who understands vibrational principles.

Expect natural cycles: Your energy and results will naturally ebb and flow throughout the 90 days. These cycles are part of the integration process, not signs of failure.

Personalize thoughtfully: While you can adjust the plan to your needs, be honest about whether changes are serving your growth or avoiding discomfort.

Focus on progress, not perfection: Celebrate improvements rather than expecting immediate mastery. Each day of practice builds your capacity, even when results aren't immediately visible.

This 90-day roadmap provides a structured approach to implementing the principles and practices from this book. By following this progressive plan, you'll develop not just intellectual understanding but embodied mastery of vibrational manifestation.

The journey doesn't end at 90 days—it simply transitions from intensive implementation to lifestyle integration. The practices that prove most valuable for you will become part of your ongoing approach to conscious creation.

Closing Note

If you've reached this final page, know that you are no longer the same person who began this journey. The science of vibration is not just a theory—it is your living reality. Manifestation is not magic; it is your birthright, grounded in laws that govern everything from atoms to emotions. My hope is that you've found more than techniques—that you have discovered the frequency of your true self, tuned your life to higher resonance, and experienced firsthand the power to create from the inside out. The journey does not end with this book; it begins with your next intention, your next vibration, your next conscious breath. Go forth and resonate—your universe is listening.

THE COMPLETE LIBRARY OF KANAV SACHDEV

Discover the Series That Calls to Your Soul

SERIES 1: THE MANI-FESTATION SERIES

The Definitive 9-Stage Blueprint for Becoming the Conscious Architect of Your Life.

- Book 1: Make the Law of Attraction Work for You Stop predicting your future and start scripting it. Go beyond passive wishing and learn the mechanical laws that govern how reality is actually built.
- Book 2: Unlocking the Power of the Subconscious Mind Your life is running on autopilot. Learn to access the hidden operating system and rewrite the deep internal programs that are silently sabotaging your success.
- Book 3: A Practical Guide to Quantum Jumping Time is an illusion. Learn to "jump" timelines and merge with the version of you who *already* has everything you desire right now.
- Book 4: Unlocking the Power of Mantra, Yantra & Rudraksha Stop doing it all with your "willpower." Activate these ancient batteries of energy—sound, geometry, and nature—to amplify your intention a thousandfold.
- Book 5: The Science of Vibrations & Frequencies Everything is energy. Learn the physics of your own aura and tune your personal frequency to the station of abundance, health, and love.
- Book 6: Shadow Work & Emotional Alchemy You cannot manifest light while ignoring the dark. Journey into your shadow to find the gold hidden in your deepest wounds and turn pain into power.
- Book 7: Sacred Rituals & Tantric Pathways Routine is mundane; ritual is magical. Fuse your intention with ancient Vedic ceremonies to turn your daily actions into unstoppable forces of nature.
- Book 8: The 40-Day Reality Shift Don't just read about change—embody it. A rigorous 40-day boot camp of affirmations and meditations to kill the old you and birth the new identity you desire.
- Book 9: Global Secrets of Creation The capstone of the series. Journey through Shamanism, Taoism, and Alchemy to discover the universal laws of creation that unite all wisdom traditions.

SERIES 2: IN SEARCH OF ZERO

A 10-Volume Initiation into the Ultimate Nature of Reality.

- Book 1: 112 Paths to Meditative Enlightenment The source code of enlightenment. 112 direct portals from Shiva to turn every breath and sensation into a doorway to the Infinite.
- Book 2: A Guide to Astral Projections Why stay locked in one body? Learn the precise mechanics of leaving the physical form to explore the universe firsthand using time-tested techniques.
- Book 3: Lucid Dreaming & the Multidimensional Mind Sleep is not just for rest; it's a laboratory. Wake up inside your dreams and use them to re-program your mind while the world sleeps.
- Book 4: 17 Vedic Meditation Techniques for Inner Stillness Modern stress requires ancient antidotes. Master the forgotten techniques of the sages that instantly soothe the nervous system and silence the noise.
- Book 5: The Akashic Mind Your soul has a memory. Access the cosmic library to heal the ancestral trauma and karmic debts that don't belong to you, consciously rewriting your destiny.
- Book 6: The Kundalini Path to Inner Alchemy You are sitting on a dormant volcano of power. Learn to safely wake the serpent energy and transform your biology from the inside out.
- Book 7: The Power of Mantra & Sound Words are spells. Discover the specific sounds (Bija Mantras) that literally rearrange the molecular structure of your consciousness and clear energetic blocks.
- Book 8: Tantra of the Void Emptiness is not nothing; it is the womb of everything. Enter the *Shunyata* and rest in the creative dark where all potential exists.
- Book 9: The Mind of Shiva Drop the ego completely. Learn to view the drama of life through the detached, blissful eyes of the Creator—the ultimate state of observer-consciousness.
- Book 10: The Elemental Path The final integration. Master Earth, Water, Fire, Air, and Space to become a living embodiment of the Cosmos.

SERIES 3: THE TANTRIC ALCHEMY SERIES

For the Seeker Ready to Walk Through Fire and Emerge Reborn.

- Book 1: The Mahāvidyā Code The Goddess is not just gentle; she is wild. Crack the code of the 10 Wisdom Goddesses to access untamed Shakti and manifest from a place of power.
- Book 2: Echoes of Mahākāla Step into the presence of Bhairava. Claim the absolute fearlessness that comes when you look death in the eye and don't blink.
- Book 3: Whispers of the Yakṣinīs Enter the 36 gates of the *Uddamareshvara Tantra*. A secret path to beauty, power, and profound inner alchemy for the dedicated adept.
- Book 4: Eye of Bhairava Transform perception itself. Based on the *Netra Tantra*, this book teaches you to dissolve illusion and see through the fire of truth.
- Book 5: The Shadow Protocol Based on the fierce *Bhuta Damara Tantra*. Confront the inner demons that haunt the subconscious and force them to serve your highest self.
- Book 6: The Sacred Tantra From chaos to wholeness. Drawing from the *Mahanirvana Tantra*, this guide heals the fragmented self in the chaos of the Kali Yuga.

SERIES 4: THE BREATH OF TIME AND DESTINY

A Cinematic Trilogy on Time, Alchemy, and Dissolution.

- Book 1: The Living Clock Shatter the tyranny of the wristwatch. Discover "Time" not as a measurement, but as a living entity you can converse with and control.
- Book 2: The Instant Oracle Stop guessing. Develop the razor-sharp intuition to know the answer before the question is even finished, accessing the wisdom of the moment.
- Book 3: The Alchemist of Breath Structure changes when the breath changes. Use the alchemy of air (Swara) to transmute the lead of your destiny into gold.

SERIES 5: THE GOD CONSCIOUSNESS

A 4-Part Journey into the Archetypes That Rule the Soul.

- Book 1: The Hanuman Consciousness Service, Strength, Devotion. Move beyond worship and inhabit the archetype of the perfect servant of the Divine.
- Book 2: The Krishna Consciousness Love, Strategy, Leela. Navigate the battlefield of life with a smile, mastering the art of divine play.
- Book 3: The Shiva Consciousness Stillness, Destruction, Presence. Anchor yourself in the unshakeable silence that witnesses the creation of worlds.
- Book 4: The Kali Consciousness The Void, The End, The Truth. Step beyond the clock and into the void of the destroyer to experience the final dissolution of the ego.

SERIES 6: MYTHOLOGY & CONSCIOUSNESS

Transmissions of Ancient Memory.

- Book 1: Katha – 108 Stories from the Vedas & Puranas Bypass the skeptical mind. Let these ancient myths act as "spiritual medicine," awakening deep memory and wisdom through the power of story.
- Book 2: Katha Shakti – 108 Stories of the Divine Mother Journey to the 109 seats of the Goddess and witness the raw, creative power of the Feminine that fuels the universe.

SERIES 7: THE NLP MASTERY SERIES

Unlock the Language of the Subconscious for Influence.

- Book 1: Unlocking Sales Success People don't buy products; they buy trust. Learn the subconscious language that makes "Yes" the only natural answer.
- Book 2: Human Potential Unlocked True leadership is resonance. Learn to lead teams not by authority, but by emotional pacing, identity design, and inner clarity.
- Book 3: The Persuasion Code Negotiation is a science. Decode the neuroscience of influence to shift beliefs and shape outcomes without force.

THE ARCHIVES OF CONSCIOUSNESS

A Roadmap for the Soul by Kanav Sachdev

You are holding one piece of a larger puzzle. This library is not a random collection of books; it is a connected system designed to guide you from where you are to where you are meant to be.

Locate your current hunger below to discover your next step.

PHASE 1: THE ARCHITECT (Master Your Reality)

For the seeker who is ready to stop waiting and start commanding. If you feel stuck, limited, or powerless, start here.

- Manifestation (Book 1): Make the Law of Attraction Work for You Stop predicting your future and start scripting it. Go beyond passive wishing and learn the mechanical laws that govern how reality is built.
- Procrastination No More: A Guide to Productivity You aren't lazy; you're just unregulated. Use neuroscience-backed strategies to break the paralysis, move from "stuck" to "unstoppable," and achieve your goals.
- Manifestation (Book 2): Unlocking the Power of the Subconscious Mind Your life is running on autopilot. Learn to access the hidden operating system and rewrite the deep internal programs that keep sabotaging your success.
- Manifestation (Book 8): The 40-Day Reality Shift Don't just read about change—embody it. A rigorous 40-day boot camp of affirmations and meditations to kill the old

you and birth the new identity you desire.

- Clockwork: Guide to Streamlined Business Operations Chaos is not a strategy. Build the systems that allow your business to run like a well-oiled machine—eliminating bottlenecks so it can grow with or without you.

PHASE 2: THE HEALER (Restore Your Soul)

For the seeker carrying heavy emotional baggage, repeating ancestral patterns, or feeling "blocked" by invisible walls.

- Manifestation (Book 6): Shadow Work & Emotional Alchemy You cannot manifest light while ignoring the dark. Journey into your shadow to find the gold hidden in your deepest wounds and turn pain into power.
- In Search of Zero (Book 5): The Akashic Mind Your soul has a memory. Access the cosmic library to heal the ancestral trauma and karmic debts that don't belong to you, consciously rewriting your destiny.
- Katha: 108 Stories from the Vedas & Puranas Bypass the skeptical mind. Let these ancient myths act as "spiritual medicine," awakening deep memory and wisdom through the power of story.
- In Search of Zero (Book 4): 17 Vedic Meditation Techniques Modern stress requires ancient antidotes. Master the forgotten techniques of the sages that instantly soothe the nervous system and silence the noise.

PHASE 3: THE MAGICIAN (Awaken Your Power)

For the seeker who wants to "turn up the voltage." Learn to use sound, energy, and vibration to bend reality.

- Manifestation (Book 5): The Science of Vibrations & Frequencies Everything is energy. Learn the physics of your own aura and tune your personal frequency to the station of abundance, health, and love.
- In Search of Zero (Book 7): The Power of Mantra &

Sound Words are spells. Discover the specific sounds (Bija Mantras) that literally rearrange the molecular structure of your consciousness and clear energetic blocks.

- Manifestation (Book 4): Mantra, Yantra & Rudraksha Stop doing it all with your "willpower." Activate these ancient batteries of energy—sound, geometry, and nature—to amplify your intention a thousandfold.
- Manifestation (Book 7): Sacred Rituals & Tantric Pathways Routine is mundane; ritual is magical. Fuse your intention with ancient Vedic ceremonies to turn your daily actions into unstoppable forces of nature.
- In Search of Zero (Book 6): The Kundalini Path to Inner Alchemy You are sitting on a dormant volcano of power. Learn to safely wake the serpent energy and transform your biology from the inside out.

PHASE 4: THE TRAVELER (Expand Your Consciousness)

For the adventurer who knows this physical world is just a tiny fraction of what exists.

- In Search of Zero (Book 2): A Guide to Astral Projections Why stay locked in one body? Learn the precise mechanics of leaving the physical form to explore the universe firsthand using time-tested techniques.
- In Search of Zero (Book 3): Lucid Dreaming & the Multidimensional Mind Sleep is not just for rest; it's a laboratory. Wake up inside your dreams and use them to reprogram your mind while the world sleeps.
- Manifestation (Book 3): A Practical Guide to Quantum Jumping Time is an illusion. Learn to "jump" timelines and merge with the version of you who already has everything you want right now.
- The Breath of Time (Book 1): The Living Clock Shatter the tyranny of the wristwatch. Discover "Time" not as a measurement, but as a living entity you can converse with and control.
- The Breath of Time (Book 2): The Instant Oracle Stop

guessing. Develop the razor-sharp intuition to know the answer before the question is even finished, accessing the wisdom of the moment.

PHASE 5: THE ALCHEMIST (Walk Through Fire)

For the advanced seeker drawn to the fierce, the hidden, and the esoteric. This path is not for the timid.

- The Tantric Alchemy Series (Book 1): The Mahāvidyā Code The Goddess is not just gentle; she is wild. Crack the code of the 10 Wisdom Goddesses to access untamed Shakti and manifest from a place of power.
- The Tantric Alchemy Series (Book 5): The Shadow Protocol Based on the fierce *Bhuta Damara Tantra*. Confront the inner demons that haunt the subconscious and force them to serve your highest self.
- The Tantric Alchemy Series (Book 2): Echoes of Mahākāla Step into the presence of Bhairava. Claim the absolute fearlessness that comes when you look death in the eye and don't blink.
- The Breath of Time (Book 3): The Alchemist of Breath Structure changes when the breath changes. Use the alchemy of air to transmute the lead of your destiny into gold.
- Katha Shakti: 108 Stories of the Divine Mother Journey to the 109 seats of the Goddess and witness the raw, creative power of the Feminine that fuels the universe.
- The Tantric Alchemy Series (Book 3): Whispers of the Yakṣinīs Enter the 36 gates of the *Uddamareshvara Tantra*. A secret path to beauty, power, and profound inner alchemy for the dedicated adept.
- The Tantric Alchemy Series (Book 4): Eye of Bhairava Transform perception itself. Based on the *Netra Tantra*, this book teaches you to dissolve illusion and see through the fire of truth.
- The Tantric Alchemy Series (Book 6): The Sacred Tantra From chaos to wholeness. Drawing from the *Mahanir-*

vana Tantra, this guide heals the fragmented self in the chaos of the Kali Yuga.

PHASE 6: THE MYSTIC (Return to Source)

The final destination. For the soul seeking silence, dissolution, and the eyes of God.

- In Search of Zero (Book 1): 112 Paths to Meditative Enlightenment The source code of enlightenment. 112 direct portals from Shiva to turn every breath and sensation into a doorway to the Infinite.
- The God Consciousness Series (Book 1): The Hanuman Consciousness Service, Strength, Devotion. Move beyond worship and inhabit the archetype of the perfect servant of the Divine.
- The God Consciousness Series (Book 2): The Krishna Consciousness Love, Strategy, Leela. Navigate the battlefield of life with a smile, mastering the art of divine play.
- The God Consciousness Series (Book 3): The Shiva Consciousness Stillness, Destruction, Presence. Anchor yourself in the unshakeable silence that witnesses the creation of worlds.
- The God Consciousness Series (Book 4): The Kali Consciousness The Void, The End, The Truth. Step beyond the clock and into the void of the destroyer to experience the final dissolution of the ego.
- In Search of Zero (Book 9): The Mind of Shiva Drop the ego completely. Learn to view the drama of life through the detached, blissful eyes of the Creator.
- In Search of Zero (Book 8): Tantra of the Void Emptiness is not nothing; it is the womb of everything. Enter the *Shunyata* and rest in the creative dark.
- In Search of Zero (Book 10): The Elemental Path The final integration. Master Earth, Water, Fire, Air, and Space to become a living embodiment of the Cosmos.

STANDALONE TITLES FOR SUCCESS

Essential Guides for the Modern Achiever.

- Procrastination No More You aren't lazy; you're just unregulated. Use neuroscience-backed strategies to break the paralysis, move from "stuck" to "unstoppable," and achieve your goals.
- Clockwork: Guide to Streamlined Business Operations Chaos is not a strategy. Build the systems that allow your business to run like a well-oiled machine—eliminating bottlenecks so it can grow with or without you.
- Global Gateway: The Complete Guide To Studying Abroad The world is your classroom. A comprehensive blueprint for the high-achiever to secure a seat at the world's top universities and launch a global career.

FOR BREAKING LAZINESS & INERTIA

The Problem: You know what to do, but you just can't get yourself to move.
The Solution: Combine *Brain Science* with *Subconscious Reprogramming*.

- Procrastination No More Stop beating yourself up. Use neuroscience to hack your dopamine reward system and trick your brain into taking action effortlessly.
- Manifestation (Book 2): Unlocking the Power of the Subconscious Mind Willpower fails; programming sticks. Rewrite the hidden script that says "I am lazy" and replace it with a drive that pulls you forward automatically.

FOR A TOTAL IDENTITY REBOOT

The Problem: You feel stuck in a version of yourself that you have outgrown. You want to change your personality, habits, and life trajectory. The Solution: The Immersion Protocol.

- Manifestation (Book 8): The 40-Day Reality Shift This is not a book; it is a death-and-rebirth ceremony. Commit to this rigorous 40-day blueprint of reprogramming to kill the old identity and permanently install the new one.

FOR CRUSHING ANXIETY & STRESS

The Problem: Your mind won't stop racing, you feel overwhelmed, or you are on the verge of burnout. The Solution: Combine Nervous System Regulation with Divine Detachment.

- In Search of Zero (Book 4): 17 Vedic Meditation Techniques Stop trying to "think" your way out of stress. Use these forgotten physiological switches to instantly calm the nervous system and silence the noise.

- The God Consciousness (Book 3): The Shiva Consciousness Stress comes from attachment. Step into the archetype of Shiva—the Witness—and anchor yourself in the unshakeable stillness that watches the chaos without being touched by it.

FOR INSTANT CLARITY & INTUITION

The Problem: You are facing a major decision, you feel confused, or you need to know what the future holds. The Solution: The Breath of Time Two-Step Protocol. (Note: This power must be unlocked sequentially. You cannot use the Oracle without the Clock.)

- Step 1: The Living Clock (Book 1) First, shatter the illusion of linear time. You must learn to feel the "heartbeat" of the Living Clock before you can read it.

- Step 2: The Instant Oracle (Book 2) Once attuned, use this guide to access the razor-sharp wisdom of the moment. Stop guessing and start "knowing" the answer before the question is even fully asked.

THE QUANTUM SHORTCUTS

Specific Tools for Immediate Shifts Sometimes you don't have time for the long road. You need a shift now. These specific combinations are "book prescriptions" designed to attack your biggest roadblocks from two angles simultaneously: the energetic and the practical.

FOR URGENT FINANCIAL GROWTH

The Problem: You need to generate wealth, close deals, or shift your financial reality quickly. The Solution: Combine Energetic Amplification with Psychological Influence.

- Manifestation (Book 4): Unlocking the Power of Mantra, Yantra & Rudraksha Money is energy, and you are currently blocked. Use ancient geometric (Yantra) and sonic (Mantra) tools to clear the path and magnetize your aura for abundance.

- NLP Series (Book 3): The Persuasion Code Once the energy is clear, you must close the deal. Use advanced neuroscience to bypass resistance, negotiate with power, and turn every conversation into an opportunity.

A PERSONAL NOTE FROM THE AUTHOR

If this book has sparked a thought, shifted your perspective, or opened a new door in your mind, I'd be deeply grateful if you could take a few minutes to leave a review.

Your words may seem small, but they carry weight. They can help someone—perhaps searching in the exact way you once were—discover this book and begin their own transformation.

Books, when written with the right intention, awaken. And your review might just be the nudge that awakens someone else.

Thank you for being part of this journey.

— Kanv Sachdev

Made in the USA
Monee, IL
12 November 2026

f4f7fe7f-7014-4f47-b6a5-394e6883a03eR01